Kitchen Mysteries

ARTS & TRADITIONS OF THE TABLE

Hervé This

Kitchen Mysteries

REVEALING THE SCIENCE OF COOKING

(Les Secrets de la Casserole)

TRANSLATED BY Jody Gladding

COLUMBIA UNIVERSITY PRESS {new york}

COLUMBIA UNIVERSITY PRESS
Publishers Since 1893
New York   Chichester, West Sussex

Translation copyright © 2007 Columbia University Press; copyright © Éditions Belin, 1993
Paperback edition 2010

This work is published with the support of the French Ministry of Culture, Centre National du Livre
(Ouvrage publié avec le concours du Ministère français chargé de la culture—Centre National du Livre)

Library of Congress Cataloging-in-Publication Data

This, Hervé.
    [Secrets de la casserole. English]
    Kitchen mysteries : revealing the science of cooking / Hervé This ; translated by Jody Gladding.
        p. cm. — (Arts and traditions of the table)
    Includes index.
    ISBN 978-0-231-14170-3 (cloth : alk. paper) — ISBN 978-0-231-14171-0 (pbk. : alk. paper) —
    ISBN 978-0-231-51203-9 (electronic)
    1. Cookery. 2. Food. 3. Gastronomy. I. Title II. Series.

TX651.T3713 2007
641.5—DC22

                                                                            2007033309

DESIGN BY VIN DANG + FERDINAND TYPE BY ISAAC TOBIN / C 10 9 8 7 6 5 4
p 10 9 8 7 6 5 4 3 2 1

Contents

———

Series Editor's Foreword

As an ever-curious if scientifically untrained food lover and sometime home sous-chef, I have long sought to grasp the rationale that science might offer to explain the not-infrequent disasters I have produced in my unscientific and amateurish cooking. Thanks to Hervé This, I discovered that I could live with the pitfalls as long as failure could become a learning experience.

What makes microwave heat so different from gas-fired ovens that it causes a *tarte Tatin* to go all soggy? Why did that soufflé fail to rise, what diabolical chemistry caused my béarnaise to liquefy? Call Hervé This. He is our necessary navigator over microwaves and through formerly murky seas of culinary science.

The chef's domain is a Faustian alchemist's laboratory. As Hervé This demonstrates charmingly and convincingly, the mysterious transformations created there rely on predictable chemical or physiological reactions that we know how to bring about, avoid, or remedy (though we be ignorant of the scientific principles behind the phenomena). But chefs and cooks, whether amateur in the home kitchen or professional behind the stoves, anyone, in short, with more than the slightest curiosity, should want to know how things work. The incontrovertible laws of chemistry and physics are here made accessible, and their practical application demonstrated.

Why is the crust of bread tastier than the crumb? Are prosciutto and other salt- or air-cured pork products really safe? How does the chemistry of salt and air "cook" a ham? Why is the decomposition brought about by the enzymes in marinades a form of chemical creativity? By what chemical process does a fine red wine become "corked"? Under a manifesto of total accessibility, Hervé This offers his reader "popular [culinary] science" in brief chatty chapters resembling sound bites from his famous TV shows. Overcoming whatever potential intimidation his use of basic scientific terms might arouse, Hervé This writes in a style both endearing and dignified, combining science, cultural history, and humor. *Kitchen Mysteries: Revealing the Science of Cooking* is a worthy companion to *Molecular Gastronomy*, the author's first book in our series Arts and Traditions of the Table.

"Molecular gastronomy": a culinary buzzword for the new millennium? In the last decades of the twentieth century controversies as to what constituted "nouvelle cuisine" or, later, "fusion cooking" inspired a veritable litany of protests from chefs denying their adhesion to such perhaps ephemeral fashions. Methinks they doth protest too much.

Today, "I don't do molecular gastronomy" has become a much intoned protest song for those reluctant to join in the avant-garde chorus led by Ferràn Adría, Pierre Gagnaire, Charlie Trotter, and Heston Blumenthal, to name but the most prominent innovators of a cuisine that some have called "cuisine scientifique" and others "deconstructive." (Mind you, Heston and the others never actually did Molecular Gastronomy, because MG is science. In the beginning, some chefs did "Molecular Cooking," and many others are still doing it, in spite of the fact that Heston and Ferran have now moved to culinary art instead of using the technological applications of MG. Deconstruction is another story, nothing to do with MG or Molecular Cooking.)

These great practitioners of what is in fact a panoply of variegated cuisines share a spirit of scientific inquiry. They are chefs who willy-nilly combine their art with science (caution: applied science cannot exist because if it is science, it's not applied, and if it is applied, it's not science but technology). No chef actually practices science, of course; they engage in a craft, because they have to produce: the kitchen is, after all, a laboratory. There, inevitably, chemical and physics experimentation is constantly under way. For example, as This emphasizes, a question ceaselessly raised and tested has to do with the ongoing importance of Maillard reactions (the action of amino acids and proteins on sugar), or what makes foods change color, to what degree does heat transform tastes or aromas? In the cooking process, a chef has to ask how to keep the primary colors of vegetables intact. What is the scientific explanation for those methods? Indeed, why do certain procedures work or inevitably fail?

Hervé This examines the why behind kitchen rituals. He makes complex science easy to grasp for nonscientists and the general reader. Here we have science and history at the service of the practical in a veritable physiology of taste, one worthy of the first promulgator of gastronomic science, Brillat-Savarin!

Albert Sonnenfeld

Kitchen Mysteries

Cooking and Science

Venial Sins, Mortal Sins

"Add the cheese béchamel to the egg whites, beaten into stiff peaks, without collapsing them!" Such vague instructions in a soufflé recipe often make amateur cooks nervous. How to avoid collapsing those laboriously beaten egg whites? In our ignorance, we begin by using what we think is a gentle—that is, slow—technique. The egg whites and the béchamel do not mix easily, so we stop before we have a homogeneous blend or we stir the two ingredients so long that the egg whites collapse. In both cases, the effect is the same: the soufflé is ruined.

Where does the fault lie? With the cookbook that takes for granted such simple techniques, known to professionals but not sufficiently mastered by the general public? With the neophyte, who naively, even presumptuously, ventures into a discipline that is not so simple as it seems?

Difficulties like those encountered in preparing a soufflé do not jeopardize our access to the realm of taste, and even the cookbook's scant instructions mark only a venial sin. With a little research, the novice will soon track down an explanation of the basic culinary techniques, and, reassured, she or he will come around to accepting—even to wishing—that cookbooks not all repeat the same advice, which he once considered them to be lacking.

On the other hand, more troubling, it seems to me, are such terse phrases as "Mix the egg yolks two by two into the cheese béchamel sauce thus prepared." Why two by two? And why not six at once if I am in a hurry? This time, the explanation is nowhere to be found. Experience alone demonstrates the validity (or not!) of the advice. A few attempts to break the rule will return the audacious cook to the wisdom of the ancients, but he will remain intellectually frustrated if he is as curious as he is epicurean.

In this work, I want to share with you the explanations that science offers for those empirical precepts handed down from chef to chef and from parent to child. Better understood, the bits of advice and suggested techniques that cookbook authors offer in passing will be better respected. Knowing the reasons behind them, you will be able to follow recipes considered difficult for their thousands of "fundamental trifles" and achieve results you never expected. You will learn to adapt recipes to the ingredients available to you; sometimes you will even modify proposed techniques according to the available utensils. Feeling equal to the task, you will be more confident and more relaxed, and you will be able to call into play all your innate creativity.

Canard à la Brillat-Savarin

To whet your appetite by giving you the chance to verify that an infusion of science can have its usefulness in cooking, I offer you a recipe that compensates for the inadequacies of the microwave: a quick *canard à l'orange*.

Who has not taken a bland, gray, tasteless piece of meat from his microwave? Should we prohibit the use of microwaves for cooking meat and restrict them to reheating prepared dishes? It would be a shame to deprive ourselves of their advantages (quick, economic, energy-efficient cooking), but we must learn the specific possibilities this new kind of cooking offers, so that we do not ask it for more than it can give. As the old, politically incorrect proverb says, even the most beautiful woman in the world can only give what she has.

Microwave cooking is no great mystery. Very simply, microwaves heat the specific parts of food that contain lots of water. In other words, if we are not careful and put a piece of meat in a microwave oven, we will only succeed in steaming it. What a shame to turn duck or beef tenderloin into stew!

Why are microwaves so deficient when used in this way? Because they skip over one of the three fundamental functions of cooking. Cooking must, of course, kill microorganisms and make tough, fibrous, or hard-to-digest foods assimilable. But it must also make food taste good.

If grilling works wonderfully, it is precisely because it fulfills these roles simultaneously. First, the surface of the meat hardens because the surface juices evaporate while the meat proteins coagulate. Second, the meat's constituents react chemically to form vividly colored molecules but also odorant or tasty ones. In other words, a flavorful and colored crust is formed. Within the piece of meat, the collagen molecules that toughen the meat are broken down.[1] The meat becomes tender. If the meat is seared, that is, cooked quite rapidly, the juices at its center do not disperse too much toward the exterior, and the meat retains its succulence and its juiciness. Biting into the meat will break the muscle fibers so that internal juices are released, bathing the mouth in a wave of delicate sensations.

Let us recognize in passing some principal chemical reactions in cooking, Maillard reactions, to which I will often return in this book: acted upon by heat, the molecules of the family to which our table sugar belongs (wrongly called carbohydrates, because these compounds are not strictly speaking made of carbon and water) and amino acids (the individual links in those large protein molecules) react and produce various odorant and tasty molecules. In cooking, this is one of the reactions we utilize to add savor even when we do not add flavorings to our dishes.

To make a *canard à l'orange* worthy of its name, the microwave will not suffice. Since microwaves heat water in particular and do not increase internal temperatures to more than 100°C (212°F, the boiling point of water) in ordinary culinary conditions, they do not promote Maillard reactions. In the *canard à la Brillat-Savarin* that I am suggesting you make, the microwave will only be used for the braising, after a quick turn in the frying pan.

Control your desire to discover this much anticipated recipe and grant me a few lines to introduce you briefly to the one to whom I have dedicated it, one of the greatest gastronomes of all time, the author of the *Treatise on the Physiology of Flavour*, which every gourmand ought to have read.[2]

---

1   Collagen is the protein that makes up collagenic tissue; it is this tissue that gives our facial skin its structure. For example, the wrinkles that we earn over the years are the result of a gradual change in the collagen. In muscles, collagenic tissue forms sheaths around cells, groups of cells, and entire muscles.

2   A hierarchy is often established between gourmands and gourmets, the latter ranked higher, in an echelon that values quality over quantity. That is a mistake. A gourmand is one who likes good food, and a gourmet is one who takes delight in wines. Thus the *gourmet-piqueurs* are professionals charged by wine merchants to identify wines produced by winegrowers.

His mother was a blue ribbon chef named Aurore (hence the name of the sauce), but Jean-Anthelme Brillat (1755–1826) took the name Savarin from one of his aunts, as a condition for becoming her heir. Because of the French Revolution, his was a turbulent career. After spending some time in exile in the United States, he returned to France, where he was named adviser to the French Supreme Court in 1800. Two years before his death, he published the book that made him famous and from which I will draw many precepts, quotations, and anecdotes in the pages that follow.

Now for that duck recipe. Begin with thighs that you have grilled in clarified butter over a very hot flame but for a very short time, long enough to allow a lovely golden crust to appear. The clarification of butter, that is, melting butter slowly and using only the liquid fatty portion of the melted product, is useful as butter thus treated does not darken during cooking. After the first grilling process, the meat is still inedible because the center remains raw, and we know that duck must be cooked! Using a paper towel, blot the fat from the surface of the thighs, and, using a syringe, inject the center of the meat with Cointreau (better yet, with Cointreau into which you have dissolved salt and infused pepper). Place the thighs in a microwave for a few minutes (the precise amount of time will vary according to the number of pieces and the power of the oven). During the cooking process, the surface of the meat will dry slightly and need no further treatment. On the other hand, the center of the meat will be "braised" in an alcohol vapor and flavored with orange (my own personal taste also prompts me to stud the flesh with cloves before microwaving it).

Spare yourself the trouble of making a sauce: it is already in the meat. No need to flambé: the alcohol has already permeated the flesh. Check your watch: you will see that putting science to work has cost you no time; quite the contrary. Furthermore, it has rejuvenated an old recipe by making it lighter.

Horresco Referens[3]

If the book you are now holding in your hands explains a few mysteries of cooking, it nevertheless sheds no light on many areas. Foods are a complex mix,

---

3   Latin for "I shudder with horror in telling it."

hard to analyze chemically. For example, Maillard reactions operate simultaneously in hundreds of compounds. The chemical combinations are countless, as are the products formed. And certain molecules present in minimal concentrations in foods perform brilliant solo parts in the grand symphony of flavors.

The natural world is so rich that cooking will always remain an art, in which work and intuition will sometimes lead to miracles. A plant like sage, for example, contains more than five hundred odorant compounds. Many a roux will thicken in our saucepans before we ever determine the exact role of these compounds in the flavor. Simple calculations show that the exploration of food combinations, compounds, and flavors will never come to an end.

So does science have no place in the kitchen? Not at all! The knowledge it produces offers simple principles that apply to the different classes of food. It explains many procedures. What you will discover here is the useful information it can provide us for eating well.

This book is not concerned with the composition of food, however. Dietary books annoy gourmands because their immediate objective is not gustatory pleasure. Often the long lists of ingredients, the tables of food constituents, in terms of fats, carbohydrates, proteins, and trace minerals, serve no purpose because they do not help answer the main question: how do the various culinary operations transform foodstuffs? How do these operations simultaneously render fibrous or indigestible foodstuffs not only assimilable but also delicious?

In one chapter of his book, Brillat-Savarin writes, "Here I meant to insert a little essay on food chemistry, and to have my readers learn into how many thousandths of carbon, hydrogen, and so forth, both they and their favorite dishes could be reduced; but I was stopped by the observation that I could hardly accomplish this except by copying the excellent chemistry books which are already in good circulation."[4] Is that really what stopped the great gastronome? Or, rather, was he applying the maxim he gives in his introduction? "I have barely touched on the many subjects which might have become dull."[5]

4 Jean Anthelme Brillat-Savarin, "Meditation 5," sec. 1 in *The Physiology of Taste; or, Meditations on Transcendental Gastronomy*, trans. M. F. K. Fisher (New York: George Macy Companies, 1949; reprint, New York: Counterpoint, 1999), part 1, p. 66.
5 Idem, author's preface, in ibid., part 1, p. 21.

Reactions in the Saucepan

Having considered what this book's subject will not be, let me move on to its central theme: science and cooking. Cooks are rarely scientists, and sometimes science frightens them. Nevertheless, the marvelous thing about science is that its subjects and its laws are simple. Notwithstanding a few explorations into the composition of matter, it asks us only to accept that our universe is composed of molecules, which in turn are composed of atoms.

That we have known since middle school. We also know that atoms are linked by chemical bonds, more or less strong according to the types of atoms. Among the atoms in a single molecule, these forces are generally strong, but between two neighboring molecules, they are weak. Often when a substance is heated moderately, only the bonds between neighboring molecules are broken. Water in the form of ice, for example, is a uniform arrangement of water molecules.

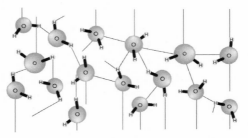

When ice is heated, the energy supplied is enough to break the bonds between the water molecules and create a liquid in which the molecules still form a co-herent mass but move in relationship to one another.

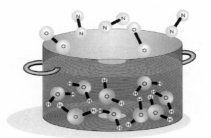

In the liquid created in this way, the molecules themselves are not transformed. The water molecules in the liquid water are identical to the water molecules in the ice. Then, as the water is heated further, it evaporates more and more, until it boils at 100°C (212°F), under ordinary pressure. The energy provided is enough to overcome the forces of cohesion binding the water molecules.

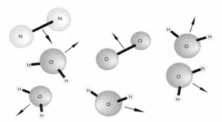

Again, however, in each molecule, the oxygen atom remains linked to two hydrogen atoms. This type of transformation is physical, not chemical, in nature. The water molecule remains a water molecule.

What the cook must keep in mind, however, is that foods are sometimes heated so much that chemical reactions can be produced as well. That is, molecules can be broken up and atoms rearranged, creating new molecules. I have already mentioned Maillard reactions, but they are not the only kind. Foods are chemical mixtures (and what is not a chemical mixture in our environment?), and the qualities we attempt to modify through cooking are manifestations of the chemical properties of these mixtures. When odorant compounds form on the surface of a roast, that is the result of a chemical reaction. When mushrooms darken after being cut, that is the result of a chemical reaction (enzymatic, but we shall return to this).

One reaction? Rather, a set of countless reactions, but we may simplify the analysis by using the biochemical classifications: carbohydrates, fats, proteins, water, mineral elements. The austerity of this decomposition allows for an overall understanding of the phenomena. Food chemistry is still in its infancy, and chemists are working hard to discover which reactions take place in foods. They are still only seeing the tip of the iceberg. We are very ignorant about the chemistry of cooking.

Universal Gastronomy

Nevertheless, there are some famous forerunners. In the mid-eighteenth century, the French cook Menon referred to the "art" of cooking, insisting on the need for experience and theory. In 1681, Denis Papin (1647–1712) invented the pressure cooker in the process of trying to discover a way to make stock from bones. The English philosopher Francis Bacon gave his life for cooking by trying to take advantage of a snowstorm to study the preservative effect of the cold. He stopped at a farm, bought a chicken, and stuffed it with snow. But he caught cold during the experiment and died of bronchitis fifteen days later.

Brillat-Savarin surveyed the scene in his time, and his admirable treatise contains a few errors that I will occasionally rectify, always paying tribute to the old master. On the other hand, I will not discuss the documents of the microbiologist Edouard de Pomiane (1875–1964). Pomiane was very popular in the 1930s, writing best sellers and creating one of the first radio programs focusing on questions of science and cooking. He believed he had invented a new science, which he called *gastrotechnie*, or gastrotechnology. This "science" encompassed nothing more than what Brillat-Savarin had already considered in his definition of "gastronomy": "Gastronomy is the intelligent knowledge of whatever concerns man's nourishment."[6] (Incidentally, it is not generally known that the word "gastronomy" comes from the title of a Greek work, *Gastronomia*, written by a contemporary of Aristotle, Archestratus, who had compiled a kind of *Michelin* guide for the ancient Mediterranean area; Joseph Berchoux [1765–1839] introduced the word into French in 1800.)

Today, the science of cooking is progressing thanks to methods of analysis perfected in the last few decades that can detect compounds present in minuscule concentrations that nevertheless play a major role in the flavor of foods. Yet it remains true that we know the temperature at the center of the planets and the sun better than the temperature at the heart of a soufflé. One of the cofounders of the scientific discipline called molecular gastronomy (the other being myself), the late Nicholas Kurti (1908–1998), a physicist at Oxford University and a member of London's very old and very respectable Royal Society,

6   Idem, "Meditation 3," sec. 18 in ibid., part 1, p. 51.

reminds us of this paradoxical fact. How to explain the paradox? I tend to think that we sometimes fear that cooking does not fall within chemistry's domain.

As proof, I offer an experiment carried out among friends, to "improve" wines. The physical chemist at Dijon's Institut National de la Recherche Agronomique (INRA), Patrick Etiévant, had discovered that two important molecules in the flavor of well-aged burgundy were paraethylphenol and para-vinylphenol. I acquired these molecules from a chemical products retailer, planning to add them to a poor-quality wine. The only comment I got from my guinea pigs was: "That smells like a chemical." An astonishing remark, because isn't everything a chemical? The foods we eat, the tools we cook with, we ourselves?

Well, it is time to discover the very substance of cooking, avoiding remarks like "it is methylmercaptan that makes urine smell after eating asparagus." What puts us off here is less the trivial nature of the remark than its useless-ness in terms of cooking. To know that asparagus contains methylmercaptan does not help us cook it. Likewise, to know that the external parts of potatoes contain alkaloids like solanine or chaconine simply allows us to eat better, not to cook better. This book aims only to promote the latter.

In this book, I examine the proven techniques, assemble the physical and chemical explanations, and do my analyses, seeking to understand without always believing that the solution given is definitive. Excuse this guide's in-adequacies if you discover any, and, through your letters, help me compile improvements for the next edition. In doing this, you will be helping all gour-mands, of which, naturally, I am one. Finally, please excuse me for sometimes being a bit academic. Like Brillat-Savarin, I am well aware that to speak without pretension and to listen with kindness, that is all that is necessary for time to flow sweetly and swiftly.

My huge regret is my inability to explain the genius of the great chefs, gifted with a sixth sense for harmonizing ingredients and creating unexpected associ-ations and surprisingly happy combinations. A veal scallop to which one adds, at the end of the cooking, a little white wine to deglaze the pan . . . and a drop of pastis? The miracle happens: a superb taste emerges. The art of cooking is not a matter of succeeding with the soufflé every time but of suspecting that pastis will transform a veal scallop. The rest is just the first course in cooking.

Mysterious as it is to many of us, this first course in cooking is indispensable if we are to devote ourselves to the study of tastes and flavors without being afraid of the béarnaise sauce turning or the soufflé collapsing on us at the last minute. When we master these things, we can follow in the footsteps of our great forebears.

The New Physiology of Flavor

The Prehistory of Tastes

Before digging into the main course—the methods of preparation—let us make a little detour useful to understanding how we eat, because we will be better cooks if we know how to distinguish the various sensations that dishes produce: tastes and flavors, colors, scents, aromas.

Aristotle knew everything, but what did he know about tastes? Let us entrust ourselves to this old philosopher. Tirelessly traversing the lyceum with his disciples, he worked up an appetite and turned his metaphysical mind toward gourmand meditations: there are "in the tastes as in the colors, on the one hand, the simple kinds which are also the opposites, that is, the sweet and the bitter; on the other hand, the kinds derived either from the first, like the unctuous, or from the second, like the salty; finally, halfway between these last two flavors, the sour, the pungent, the astringent, and the acid, more or less; these seem to be, in effect, the different tastes."

Aristotle is not the only authority to have appreciated oral sensations. In particular, in the eighteenth century the great Linnaeus also applied his talents to tastes, but paradoxically the most famous of systematicians, the father of botanical classification, lacked some systematic spirit, because he mixed together the moist, the dry, the acid, the bitter, the fatty, the astringent, the sweet, the sour, the viscous, the salty. He put them all pell-mell in the same bag for us, this mix of tastes and mechanical sensations.

A Frenchman deserves the credit for establishing a little order in the domain of oral impressions. In 1824 the great chemist Michel-Eugene Chevreul

(1786–1889), famous especially for his work on fats, distinguished the olfactory, gustatory, and tactile sensations. He recognized that the perception of hot or cold is distinct from that of sweet or bitter. He separated out the tactile sensations of the oral cavity, as well as the proprioceptive sensations (for example, toughness). With Chevreul, the taste of physiologists—one component of flavor—was distinguished from everyday sensation, where all the sensations associated with the absorption of food and drink are mixed.

In the same period but in a different circle, among the gourmands centered around Brillat-Savarin, the only confusion that continued to reign was between tastes and smells. The tongue was known to perceive tastes, but the nose was also believed to be a receptor. Apart from a few more or less harmless errors, the remarks made in the *Treatise on the Physiology of Taste* are as insightful as their author is passionate about cooking: "The number of tastes is infinite, since every soluble body has a special flavor which does not wholly resemble any other. . . . Up to the present time there is not a single circumstance in which a given taste has been analyzed with stern exactitude, so that we have been forced to depend on a small number of generalizations such as *sweet, sugary, sour, bitter*, and other like ones which express, in the end, no more than the words *agreeable* or *disagreeable*."[1] On the other hand, a bit later, Brillat-Savarin adds that "any sapid substance is perforce odorant."[2] He had forgotten that some molecules that are hardly volatile at all at ambient temperatures and thus odorless nevertheless bind easily to taste receptors on the tongue and palate and therefore have a taste. Salt, for example, is sapid but odorless.

Modern Meanderings and Recent Revelations

In trying to learn how we perceive food, physiologists first discovered the taste buds, that is, groups of sensitive cells that are responsible for detecting tasty—or sapid—molecules. In all mammals, taste is ensured by these receptors, distributed throughout the mouth, on the palate, epiglottis, pharynx, and especially

---

1   Jean Anthelme Brillat-Savarin, "Meditation 2," sec. 9 in *The Physiology of Taste; or, Meditations on Transcendental Gastronomy*, trans. M. F. K. Fisher (New York: George Macy Companies, 1949; reprint, New York: Counterpoint, 1999), part 1, p. 38.
2   Idem, "Meditation 2," sec. 10, in ibid., part 1, p. 39. A food that has a taste is "sapid."

the tongue. Our tongues have about nine thousand taste buds, in groups of fifty to one hundred, loaded with nerve endings. The number of taste buds seems to diminish with age, especially after the age of forty-five.[3]

Classical works have been reexamined. The alchemists said this of taste and smell: *corpora non agunt nisi soluta* (bodies are only capable of action in the divided state). They thought in macroscopic terms: nutmeg only has a flavor when reduced to a powder. In microscopic terms, alchemical law must be articulated in this way: a molecule is only sapid if it is soluble in water and has one or many receptors. If it is soluble in water, it is circulated through the saliva to the "nervous and sensitive tufts," as Alexandre Balthazar Laurent Grimod de la Reynière called the taste buds.[4] Receptors are needed to induce some sensation, however: water-soluble molecules that lack receptors do not deliver taste.

Petroleum jelly has no taste because its compounds do not dissolve in saliva. Apparently, sapidity results from the establishment of bonds between the sapid molecules and taste bud receptors. A molecule only has a taste if it is linked to the receptors present on the surface of the gustatory cells in the mouth. This connection seems to take place through a lock-and-key system. Because of complementary forms or electrical charges, the sapid molecule links to its specific receptor molecule and stimulates the nerves that relay the perception of a taste to the brain. The weakness of these links has the advantage of letting us sense different flavors at short intervals. One taste dispels another.

We can also understand why our forebears had so much difficulty distinguishing tastes, odors, and proprioceptive sensations. These various perceptions are relayed along nerve pathways that merge upon entering the brain. The perception of a scent can alter our perception of a taste, for example. The flavor of a dish can depend on its temperature.

---

3 Recent discoveries indicate, however, that this may not be the reason that older people lose their sense of taste. Reduced consistency perception, resulting from dental replacements, seems the more likely cause.

4 Grimod de la Reynière (1758–1838), the father of gourmand literature at the time of the Revolution, was the son of a farmer general. He is known especially for his *Gourmands' Almanac*, in which he proposes in particular a culinary tour through Paris, and for his *Manual for Hosts*, in which he explains how to cut meats, compose a menu, conduct oneself as a polite host, and, more generally, behave at the table. His remarks are not outdated.

To study the perception of pure tastes, sensorial physiologists today use standardized experimental protocols and devices that gently blow air into the noses of the subjects being tested. If odors no longer pass through the retronasal openings (connecting the mouth and nose), the subjects perceive the true taste of the foods, the quintessential sapidity, as it were.

Despite the incontrovertible results recently obtained, the public and even certain distinguished scientists still believe that there are only four tastes. The error dates back to 1916, when the chemist Hans Henning proposed his "theory of the localization of receptors," according to which the mouth supposedly perceived only four tastes (salty, sour, sweet, bitter), through specialized taste buds confined to certain regions of the tongue. Sweet was supposedly perceived by taste buds located on the tip of the tongue, bitter by taste buds at the base of the tongue, salty by the front edges, and sour by the back edges.

Recent physiological analyses have revealed how wrong this theory is. First of all, even though the salt receptors are more numerous along the front edges of the tongue, they are present throughout the mouth and all over the tongue. Similarly, the sweet, sour, and bitter receptors are present throughout, though in varying proportions. Furthermore, licorice, for example, because of glycyrrhizic acid, is neither sweet, nor bitter, nor salty, nor sour. And the molecules acting as gustatory receptors seem very much more varied than once supposed, forming weak bonds with molecules sometimes very different from one another.

Recent studies have not called into question the reality of the salty taste, which is actually due only to sodium ions, or the sour taste, which is due to hydrogen ions, but they have demonstrated the vast and varied nature of the domain of taste and confirmed Brillat-Savarin's vision.[5] Salt forms molecular structures with food proteins, structures that are stable in the cold but destroyed by heat. Salt having thus formed what chemists call a complex cannot stimulate the taste buds. That is why a given amount of salt by itself produces a salty taste, cold, and that is also why, given equal concentrations of salt, raw products seem less salty than warm, cooked products.

---

5   Sodium and hydrogen ions are sodium and hydrogen atoms that have lost an electron. The chloride ion, the partner of the sodium ion in table salt, acts principally to stimulate the receptors.

In addition, fat often does not seem salty, because it does not dissolve salt and contains little water, which does. On the other hand, it is good at dissolving many odorant molecules. It is primarily the fat in meat that gives it its characteristic flavor. Try the experiment of cooking a lean piece of pork with lamb fat. Say nothing to your dinner guests and ask them what they think they are eating.

Licorice, with its glycyrrhizic acid, is not the only substance with a taste that does not appear in the list sanctioned by ignorance. Japanese physiologists have demonstrated the need to add the taste umami as well. Umami is said to be a universal taste, although this is a complicated matter. The fact is that umami characterizes the taste of broths called dashi, made by infusing kombu (kelp) in hot water. During the infusion process, primarily two amino acids are released, glutamic acid and alanine, so that, strictly speaking, umami is the taste of the combination of these two products, not of glutamic acid, as has been claimed. So, are there four tastes, or five, or six? None of the above. Numerous molecules, the various amino acids or quinine (the prototypical bitter molecule), for instance, among many others, have unique tastes that cannot be reduced to a combination of other tastes.

Even sweet is more complex than we once imagined. The various modern sweeteners sweeten everything, but they do not all have the same sweet taste. As for the relationships between sweet and bitter, they are astonishing. Certain molecules, such as methylmannopyranoside, have both a sweet and bitter taste, or only sweet, or only bitter, depending on the individual. Why? We do not know, but recent scientific works provide glimpses of new phenomena.

I would like to propose a little detour (another one, my dear gastronomads) into two of these studies, one on sweet tastes, and one on strange molecules in the form of an L that are simultaneously bitter and sweet.

Recent Progress in the Chemistry of Sweeteners

Since scientific studies of the gustatory receptors are difficult because these receptors have only a weak affinity for the sapid molecules, certain physiologists are analyzing gustatory phenomena indirectly by having subjects taste various sweet molecules, for example, every day for several months. In a laboratory in

Massy, near Paris, many hundreds of individuals tested twenty sapid molecules in this way, using the nasal air current device previously described.

Thus, in the early 1980s, the neurophysiologist Annick Faurion and her colleagues discovered that the threshold for detecting sucrose, that is, the smallest quantity of table sugar perceptible in a fixed quantity of water, varies from one individual to another. Likewise, the thresholds for perceiving various other sweeteners are specific to individuals. In other words, the quantity of sugar that we take in our coffee depends not only on the sensation we like to have but also on our own personal sensitivity to the sweetening molecule. Moreover, the sensitivity threshold depends on the sweetening molecules themselves. Some individuals are more sensitive to sucrose (table sugar), others to glucose (the sugar in honey or grapes).

What is fascinating, though not surprising, is that the detection thresholds evolve through "learning." Over the course of the trials, the thresholds decreased; that is to say, sensitivity increased. Furthermore, when the training for one molecule ended, that is, when the detection threshold ceased to vary, it continued for other molecules. What a stroke of luck! This kind of observation demonstrates that, if we want to, we can train ourselves to develop fine palates.

Finally, comparisons among various molecules in different concentrations revealed an additional complexity in the gustatory system. The sweet taste of a sweetening molecule depends on its concentration. That is an effect we must consider, looking toward the day when we reach the final stage of mastering the slightest variation in flavor in the dishes we prepare.

What relationship exists between a molecule's structure and its taste? The indirect studies done in Massy and elsewhere have not answered this question, of interest to higher gastronomy. Yet if we understood these structure-activity relationships, as scientists call them, we could synthesize molecules tailor-made to individual tastes!

Because of the huge market for synthetic sweeteners, this subject has been addressed with special attention to sweet molecules, and enticing prospects appeared when Murray Goodman and his colleagues at the University of San Diego tested subjects with peptidic sweeteners (peptides are small molecules formed from chains of only a few amino acids). As in many artificial sweeteners—aspartame, for example—these molecules contain two rings of atoms, only the first of which can bond to water molecules, linked by a short chain of atoms in the form of an elbow at a right angle. The rings are semicoplanar, and the complete molecule forms an L-shape.

By altering such molecules so that the two rings are no longer coplanar, the San Diego chemists first obtained molecules with no taste. Then, by placing one flexible molecular part between the rings, they created molecules in which the rings could turn in relationship to one another. The continuous movements of the molecules can actually make the rings turn incessantly and very rapidly, at a speed that varies according to the relative orientation of the rings.

The taste of these molecules is . . . unpredictable. Some seem bitter at first, then sweet, whereas others are initially sweet, then bitter. This strange property might result from certain molecules being in a sweet configuration longer and bonding initially to the sweet receptors, while others, in a bitter configuration longer, bond more to the bitter receptors. When are we going to have the same "blinking" effect with other tastes?

We have not heard the final word in this gustatory adventure. As Brillat-Savarin sensed, tastes are astonishingly complex. Even without considering the "flashing light" tastes discussed above, studies seem to indicate that tastes inhabit a ten-dimensional space. In other words, tastes seem to be infinite in number, and ten descriptors at least would be necessary to talk about them. We are falling far short of the mark with only sour, bitter, sweet, and salty.

Does Taste Lose Its Edge As One Eats?

Do we perceive the taste of a dish or a drink less well after consuming a great deal of it? This question deserves study, because Brillat-Savarin affirmed—with as much authority as good reason, I believe—that "the most delicious rarity loses its influence when its quantity is stingy."[6] Yet what would be the interest in consuming a dish in abundance if the perception that we have of it and the pleasure that it provides us disappear after a few mouthfuls?

Let me pose the question in concrete terms: does the taste of mustard disappear when we overuse this condiment? Do we lose our sensitivity to wine when

6   Brillat-Savarin, "Meditation 13," sec. 69 in *The Physiology of Taste*, part 1, p. 177.

we allow ourselves the time to taste it and to examine all the components of its bouquet? Or, on the contrary, does practice in the perception of flavor increase sensitivity through the phenomenon of training?

Let me clarify these ideas by stating first of all that the term "fatigue" can be used in several ways. The first is an alteration in the physiological state of the muscles, which takes place only in very rare cases when we eat tough or hard products. The second corresponds to a progressive incapacity of the nervous system to analyze the signals it receives. This is the mental fatigue brought on by psychomotor (for example, dactylography) or intellectual (the case of flight controllers) tasks. If we acknowledge that the exercise of gustatory or olfactory perception is a recognition of forms, like dactylography or flight control, we can assume that mental fatigue can also occur in the sensory evaluation of food products.

Third, we also call a waning of interest in what we are doing "fatigue," because the activity is monotonous or because we consider it too difficult. This form of fatigue should be called "lassitude" instead. Such fatigue does not seem to apply to gourmands. How can they get tired of good things?

Finally, fatigue can be the weakening of a sensation as a result of constant exposure to a stimulus. We no longer notice the odor of a stuffy room a few minutes after entering it. This phenomenon is an inevitable adaptation, but since it occurs as much at the beginning as at the end of the tasting process, it seems wrong to consider it a true instance of fatigue. Furthermore, it may be that an adaptation to stimuli increases the quality of perception. Wine tasters rinse their mouths with wine (thus adapting their palates) before beginning an evaluation session, to provide themselves with a point of reference, just as musicians tune their instruments together before a concert. This phenomenon is well known among taste physiologists, who have observed that the threshold of perception for sucrose (table sugar) in water is lower (one is more sensitive to it) when subjects rinse their mouths with a sucrose solution before the trials than when they do not rinse or rinse with plain water.

To learn definitively whether the taste sense is dulled or not, François Sauvageot and his colleagues at Dijon University did sensory evaluation tests on subjects in which the difficulty of the task proposed to any subject at a given time depended on the quality of the prior response. When subjects gave correct responses, the next tests they had to pass were more complicated. When they made mistakes, the next tests were easier. The trials lasted four to five hours, with a thirty-minute break midway through the session. On average, the results for tasting did not deteriorate over the course of the sessions. Taste was not dulled. Good news for gourmands, who, without knowing the results of these experiments, must have hoped for this outcome.

Taste and Colors

It is sometimes said that the colors on a table are half the meal. No doubt that is true, in a sense: we do not expect the same kind of pleasure when entering a dining room glimmering with candles, crystal, and silverware as when we sit down to eat at a counter covered with an oilcloth in garish hues.

Do colors determine the flavor of a dish in the same way the temperature of a food alters its taste? It is hard to answer that question because oral pleasure can never be reduced to a single factor. Since gastronomy is precisely the art of combining pleasures, it would go very much against the grain to isolate out colors in order to examine their hedonic power.

So let us concentrate instead on the strange relationship that seems to exist between the color of a dish and the hunger it prompts. Intuitively, cooks strive to retain the fresh color of vegetables, a certain pinkness to meats, the white of fish. Pastry cooks have the time of their lives creating creams in tempting colors.

A classic cookbook published in the 1960s by the French food critic Curnonsky introduced the prize-winning creations of pastry cooks who used methylene blue to color their cakes.[7] Cooks rarely resort to such colorations, but they know that gray meat or yellowish leeks are not appealing.

In his *Grand dictionnaire de cuisine*, Alexandre Dumas lists several "inoffensive" food colorings that can brighten dishes:

---

7   The "elected prince of gastronomes," Maurice Edmond Sailland (1872 [Angers]–1956 [Paris]) was a writer and journalist. His pen name originated with Alphonse Allais, who suggested to him "pourquoi pas sky"? (why not "sky"?)—this being a time of Franco-Russian friendship—which became in Latin "*cur non sky.*"

BLUE: indigo diluted in water

YELLOW: gamboge or saffron

GREEN: juice of spinach leaves or crushed green wheat, cooked over a flame,
            strained, and diluted in sugar water

RED: cochineal and alum powder boiled in water

CRIMSON: pollen from dried wild carrot flowers diluted in water or elderberry juice
            diluted in water

PURPLE: cochineal and Prussian blue

ORANGE: saffron and cochineal

Are these colorings really inoffensive? And, more important, are they enticing? The following anecdote demonstrates that Curnonsky hit the nail on
the head when he remarked that "things are good when they have the flavor
[and the color, let us add] of what they are." For a dinner that later became
famous, the amphitryon had wanted all the dishes to be green, as well as all
the objects on the table and in the dining room: tablecloth, napkins, place
settings. The guests had a very hard time swallowing even a few mouthfuls,
and some departed, leaving their host to clean up the little they had temporarily
forced down. More recently, in taste trials, even competent judges have mistaken orange juice dyed blue for blueberry juice or even white wine, colored
with flavorless pigments, for red. Let us not force the hand of nature but rather
compensate when we degrade it. We can certainly give wilted vegetables their
colors back, but why not eat fresh or perfectly cooked ones?

How Do We Avoid Undesirable Darkening?

Without some precautionary care, mushrooms that have been sliced too soon
before a meal do little honor to the table where they are served. They go black,
as though mourning their freshness. Similarly, who wants to eat bananas, apricots, cherries, potatoes, apples, peaches, or avocados that were cut too soon and
then left out?

These color changes occur because all fruits and vegetables contain active
molecules, called enzymes, that chemically transform the phenolic compounds

they also contain into brown or gray products.[8] When a fruit or vegetable is cut, the cells along that cut are broken, discharging the phenolics and the blackening enzymes to the cut surface. In the presence of oxygen from the air, the enzymes do their unfortunate work.

How to avoid such darkening? By inhibiting or destroying the freed enzymes. Have you noticed that lemons, melons, and tomatoes escape the strict law of darkening? That is because their natural acidity—caused notably by ascorbic acid—and the presence of vitamin C block the enzymes. Likewise, vinegar, which is even more acidic, preserves pickles, capers, and *cristes-marines* (the fleshy leaves of a carrot family plant that grows in sand on the French coasts).

Cooling and cooking have the same effect. Cooling slows down the oxidation process (cooling by ten degrees cuts by half the speed of the enzyme action). Cooking "denatures" the enzymes, which are proteins, that is, long linear molecules folded back on themselves in a specific way by weak chemical bonds. This folding gives them their functional properties, but heat suppresses their activity by breaking the weak bonds. Thus these molecular balls of yarn are inactivated.

Salt blocks enzymes as well; often what we put in condiments helps with preservation. And finally, vitamin C slows down the work of enzymes. In his work entitled *On Food and Cooking*, Harold McGee reports how this important vitamin was isolated, moreover, thanks to this peculiarity. Around 1925, the Hungarian chemist Albert Szent-Györgyi became interested in the chemistry of vegetables because he had observed a resemblance between the way damaged fruit turns brown and a disease of the adrenal glands in humans. He analyzed the vegetables that did not turn brown and noted that their juice slowed down the darkening of other vegetables. Isolating the active substance, he discovered that it was an acid, which he named ignosic acid. It was vitamin C, a compound indispensable to life.

---

8 A phenol is a molecule formed from six carbon (c) atoms in a hexagonal ring, all but one of them linked to a hydrogen molecule (H); the last atom is linked to an alcohol group, that is, an oxygen atom (o) linked to a hydrogen atom (H). Phenolic compounds are larger molecules that include at least one of these structures.

How Did the Salmon Make the Lobster Blush?

The saga of food colors goes on forever. But I will close the discussion here with a curiosity. Why do crab, shrimp, crayfish, and lobster turn red when they are scalded?

This is no great mystery. The shells of crustaceans contain a molecule with four oxygen molecules, astaxanthin, the color of which does not appear in living animals because the molecules are linked to proteins and thus form a dark blue complex. Cooking sea creatures breaks up this complex (as in the case of enzymes, the weak chemical bonds are broken), and the red color of astaxanthin appears. In salmon, astaxanthin is naturally present in its dissociated form, hence the pleasing pink flesh of specimens both living and dead, but here the color is also due to other molecules that are cousins of the orange carotene molecule present in carrots.

Scents and Odors

How sad for those who perceive tastes and colors but not odors! Because our noses are very important for detecting flavor. We say of a gourmand that he has a fine palate, but we should say, as in perfumery, that he has a great nose (or maybe that he is a Cyrano?).

Let me first touch on semantics. Odor is odor, that is, what we feel with our olfactory system. Aroma is the odor of a plant that has some odor, just as bouquet is the odor of wine specifically.

How can a cook orchestrate odors? Even though, in cooking jargon, a chef's stove is referred to as his piano, chefs are more organists than pianists. They must play many registers at the same time, and each register must produce its own harmony, harmonized to the harmony of the other registers. I make no claims here to offering a recipe for virtuosity in a few lines, only to providing a glimpse of the paths that lead toward better cooking.

A harmony of odors is not the easiest to achieve, but it is what we perceive first, along with colors, and perhaps with even greater intensity. The guests are not yet seated at the table when their own scents have already mixed with those of the hearth, the candles shedding their warm and flickering glow. The door to the dining room opens, the first dish arrives, it is uncovered, and its odors are released. How to make this important moment a success?

How Can We Use Odorant Molecules?

The answer to this question is, with caution! Odorant molecules are generally fragile and volatile, because they are organic. A little too much heat, and the odors so skillfully blended escape the dish or are degraded into potentially acrid or bitter molecules. Thus the rule should be to limit the temperature or to plan on the addition or creation of odors at the end of the cooking process. Pepper, for example, cannot be cooked too long before becoming acrid; parsley, too, must be added at the end of cooking.

So, how to use odorant molecules? Cautiously, as I have said. Like the active constituents in medicines, these molecules are so powerful that the prescribed dose must not be exceeded, no matter what. The powder obtained by grating a whole nutmeg contains enough toxic molecules to kill the most robust among us. In concentrations greater than four parts per million, paraethylphenol no longer gives burgundy wines their bouquet of old leather but an unpleasant chemical odor.

How to use odorant molecules? With caution but also with discernment. It is necessary to know that odorant molecules, generally organic, dissolve easily in organic solvents but sometimes not in water. In meat, for example, as I have noted, it is the fat in particular that contains the odorant molecules, because the flesh is all water, whereas the fat is all organic.

Let me explain. Molecules dissolve in a certain medium because they establish so-called weak chemical bonds with the molecules of that medium. These bonds have about the same strength as the ones that link the molecules of a liquid to one another and keep them from volatilizing too quickly in the ambient temperature, as their natural agitation would prompt them to do. They are much weaker than those that make cooking salt a solid, composed of a regular network in which the sodium atoms (Na) alternate with the chlorine atoms (Cl).

In salt, the sodium atoms have yielded one of their electrons to the chlorine atoms. Since opposite electrical charges attract, a bit like magnets, the chlorine and sodium atoms are strongly bonded and form a solid. In water, on the other hand, the water molecules, composed of one oxygen molecule (O) and two hydrogen molecules (H), form structures that are not inclined to exchange electrons. The chemical bonds are within the water molecules, not between them.

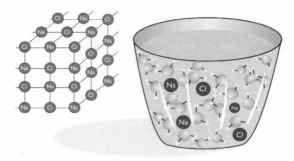

So why do water molecules combine in a liquid (water) that only boils at a fairly high temperature? Because the oxygen atoms are partial to electrons. They share them with the hydrogen atoms to build the molecule, but they keep the lion's share for themselves. On the whole, the water molecule has a slight electron imbalance on the side of the oxygen. The negative charges of the oxygen in a water molecule attract the positive charges of the hydrogen atom in another water molecule, thus linking the molecules together. This bond is weaker than an intramolecular chemical bond and is called a hydrogen bond.

And what does any of that have to do with odorant molecules? Well, molecules are often constituted with an electronic equilibrium, not possessing many atoms eager for electrons. Thus many mint- or resin-scented molecules are terpenes, that is, molecular structures that result from assembling many isoprene patterns (that is, a chemical group of five carbon atoms in the form of a Y, with three carbon atoms on the central trunk and one carbon atom at the end of each branch). This pattern is not established through the hydrogen bond, and for this reason terpenes are insoluble in water. Better yet, they are excluded from it, in the same way that fat is not miscible in water, since birds of a feather flock together. Water stays with water and excludes molecules that do not share with it the possibility of establishing a hydrogen bond (ethylic alcohol, on the other hand, includes an oxygen atom that allows the hydrogen bond to be established, thus alcohol is soluble in water).

In short, terpenic odorant molecules do not dissolve well in flesh, which is primarily an aqueous medium (the cells of the flesh are full of water); they are primarily distributed in fat. I have already indicated this, but the precept is so important that it bears repeating: fat brings out the features of the meat with which it is associated. It is mutton fat that gives mutton its taste and odor, beef fat that gives taste and odor to beef—to such an extent that, by cooking a lean beef filet in duck fat, one can make a sort of culinary chimera, a cross between beef and duck. Here we also find the explanation of the old saying "fat is good." And for good reason! It is what gives food its flavor, plain and simple. What is true of terpenes, the molecules of sensuality, widely used by the perfume industry, is equally true for the alkanes and alkenes, two classes of molecules consisting only of carbon and hydrogen atoms that often produce slightly fruity odors, as well as for many organic molecules, including the ones that plants and animals store and synthesize. It is the cook's duty to respect them, to use them with care, and to consider carefully the phase—fat or water—where they will tend to be distributed.

This introduction explains the advantages of cooking in parchment paper or braising. The temperature does not exceed the boiling point of water, so the aromas do not deteriorate; what is more, they are trapped and recycled in the food from which they would otherwise tend to escape. This is also why the new technique of cooking in a vacuum at low temperatures is a blessed gift to gourmands. After rapid roasting and an injection of aromatics, the foods are enclosed in a plastic bag from which the air is removed. Then cooking takes place at temperatures as low as 60°C (140°F). The proteins coagulate, which is the mark of all cooking processes, but the collagenic tissue does not contract too much. The juice remains in the food, which thus retains its succulence. As a bonus, the odorant molecules remain in the food because they are not expelled by the heat. Tender and flavorful, meats prepared using this technique make my mouth water at the mere mention of them.

A word, finally, on the odorant molecules that escape during cooking. Open the kitchen door while the cook is officiating, and the dish will come to your nostrils. A different dish from the one being served, however, because the warmed organic molecules tend to react with the oxygen in the air. Oxygen is very aggressive; it rusts iron, after all. In short, it is no longer the odor of the dish that we perceive but a complex mix of molecules more or less derived from it. That is why it is often preferable to cook with the door closed. To the surprise you've kept in store for your guests, you add the courtesy of not subjecting their noses to the odorant residues of the preparation.

Spice or Aromatic?

Is saffron a spice or aromatic? It has a smell but is not pungent. It serves to enhance the fragrance of a dish (from the Latin *fragare*, to smell) by providing odorant molecules. It is an aromatic. Is pepper a spice or aromatic? It awakens the flavor, but its odor is not the main reason cooks use it. It is a spice.

Spices add a little mischief to dishes; aromatics serve to revive memories, like Proust's famous madeleine, which caused him to relive his childhood at his maternal grandmother's (like all olfactory signals, scents are processed by the brain's limbic system, which also manages memories and emotions).

Distinguishing spices from aromatics is an exercise to which all cooks must devote themselves in order to master their art. It is not an easy exercise. Garlic, for example, is pungent and scented; it serves to awaken the flavor and to enhance the fragrance of a dish. It is both a spice and an aromatic.

How would you like to try classifying the aromatic supplements used in cooking? It is up to you to answer the question "spice or aromatic" for cinnamon, sesame, watercress, radish, anise, dill, coriander, cumin, fennel, thyme, basil, sage, rosemary, mint, marjoram, onion, parsley, lupin, cardamon, oregano, bay leaf, chive, absinthe, leek, pimento, mustard, caraway, celery, sugar, honey, vinegar, savory, juniper, ginger, capers, olives, chervil, burnet, nutmeg, sorrel, tarragon, myrtle, horseradish, wild celery, black cumin, purslane, nard, rue, malaguetta pepper, garum, lovage, star anise, masterwort, hyssop, mace, pennyroyal . . .

Why Does Bread Crust Have More Flavor than the Crumb?

Why does bread crust have more flavor than the crumb? Why must meats be seared in butter when preparing a stock for a *sauce espagnole*, for example? Why must a leg of lamb be rubbed with oil before it is put in the oven? Why is some beer golden? Why do roasted coffee and chocolate smell so good?

There are countless questions of this kind in cooking, but the answer to many would be, in short, "Maillard reactions." Indeed, it is these chemical reactions that create brown, odorant, and sapid compounds in cooking.

Universal as these famous, oft-mentioned Maillard reactions are, they are still not well known. Nevertheless, the principle is a simple one: as soon as molecules containing a chemical amine group NH_2 (one atom of nitrogen—N—linked to two atoms of hydrogen—H)—like the amino acids in all proteins, for example—are heated in the presence of some specific but common sugar such as glucose, a water molecule is eliminated and the two reagents are bonded in what is called a Schiff base. Let us not linger over this compound, since it is more or less rapidly replaced by an Amadori product, which will react with other compounds to form cyclical "aromatic" molecules.

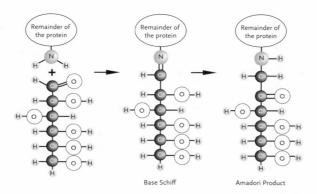

Base Schiff · Amadori Product

As their name indicates, these aromatic rings give odorant properties to the compounds that contain them; some of them also have a strong color.

Whole books of scientific articles on Maillard reactions are published with some regularity, and in 1990 a famous chemistry journal devoted a comprehensive article more than twenty pages long to this topic, describing many of the odors created. Nevertheless, the products of Maillard reactions are countless and still not sufficiently well known. The golden brown that cooks look for when sautéing in fats is produced by many reactions, but Maillard reactions are preeminent. The reactions take place at the high temperatures attained by the fat, whereas they do not occur when foods are boiled, since the temperature is limited to the boiling point of water: 100°C (212°F).

How to improve our cooking now that we know the power of Maillard reactions? By using them! As we cook, let us watch for sugar-protein combinations. Let us think of Peking duck. Let us heat foods quickly at first, so that Maillard reactions can take place, then more cautiously, in such a way that the actual cooking takes place without eliminating volatile compounds. Do we want to cook meat in a microwave? Let us not forget to sear the meat before cooking it, and let us oil or butter the surfaces so that the heat is transmitted to them efficiently, a subject to which I will return in the chapter on cooking (see page 58).

Finally, a Few Curiosities

Let us not leave the realm of odors without discovering a few surprising, even revelatory, curiosities.

First of all, the scent of oranges is primarily due to a terpene, limonene, a molecule that is the mirror image of the molecule that contributes to the scent of lemons. The shared acidity of these two fruits is due to citric acid (designated by E 330 on food labels), and their orange color comes from carotenoids, which give carrots their color. Now, carotenoids are also present in grass. When grass is cut and dries in the sun, the carotenoid molecules decompose into ionone molecules, which smell like hay ($C_{13}H_{20}O$), a compound of essential oil of violet.

Moral: the tastes and the smells that interact when we enjoy a dish result from the dynamic and complex coexistence of sapid molecules and color molecules, very difficult to classify and master. The art of cooking lies in recognizing their good effects and harmonizing them. Think of that organ. . . .

Soup

Why Does Soup Cool Down When We Blow on It?

Those of you who love to eat know why cooking inspires enthusiasm. Do you know why science has its enthusiasts as well? Because it holds the key to a secret world, parallel to the one we all know but so different! Some, who pass for gentle dreamers among their more businesslike friends, have gotten lost there.

What is this world made of? Atoms and molecules in constant motion. What do these molecules look like? That all depends on who imagines them and the framework in which they are placed. To the young chemist's eyes, they are often groupings of colored spheres that form or disintegrate in flasks, beakers, test tubes . . . or soup pots. For physicists, they are sometimes hard, compact spheres that bounce off one another like billiard balls, sometimes small solar systems that attract each other like magnets, still other times ripples analogous to those that spread across the surface of a lake.

And what about soup, an assemblage of molecules like all other matter in the universe, what image can we conjure for it? The gourmand, cheerfully attentive to the sensations he is about to discover, finds a world of odors there, the assurance of a thirst satisfied. We invite him to follow us to the molecular world to discover another facet of the subject about which a great cook like Marie-Antoine Carême (1783–1833) wrote an entire book.

The soup we are going to explore is very hot, and rising slowly from it are wisps of odorant vapors. For physicists, the soup is behaving like water. That helps us simplify the problem. The scene is as follows: liquid water in the soup tureen; air above it; vapor rising into the air. Let us begin with the vapor, which is composed of water that has become gaseous. Heated, the water molecules move so rapidly that they overcome the forces holding them in the liquid and escape into the air, where they are gradually absorbed. Lighter than the air, the warmth rises just like a piece of wood in water or a hot air balloon, since the same phenomenon is at work in all three examples. (If we see some white smoke over the soup, it is because the water molecules in air can mix locally and make microscopic water droplets: indeed, we see a smoke, not vapor.)

On the other hand, the water in the soup is composed of countless molecules moving at various speeds but generally without enough energy to escape the liquid. Moving very rapidly, some do escape, but in striking the air over the soup they are sent back into the liquid. Others, moving more slowly, are only able to drift about in the liquid, zig-zagging and haphazardly colliding with neighboring molecules.

In the air, the molecules of nitrogen, oxygen, and various other chemical elements are also moving about and will sometimes strike violently against the water molecules in the soup, dislodging them as in a boules game. Following shock after shock, an equilibrium is established between the soup and the layer of air immediately above it. The air temperature and the soup temperature become the same.

If we blow on the soup, the air above it, charged with evaporated air molecules, is replaced by dry air. Thus the evaporated molecules cannot reenter the soup. Then, other molecules from the soup evaporate, are taken away by the blown air, and so on. By blowing, we encourage the evaporation of the soup.

Since the molecules that evaporate are precisely the ones with the most energy, only the molecules with the least energy remain in the soup. Thus evaporation corresponds to a decrease in the energy of the liquid, that is to say, a cooling. In other words, to cool, blow. The phenomenon is the same as the one you experience when you emerge from a dip in the ocean on a windy day. By evaporating the water that remains on your skin, the wind cools you.

A word of caution, all the same: if the soup is thick, you will have to stir it as you blow on it in order to cool the entire bowlful. Otherwise, only the surface will be cooled. The soup's viscosity will prevent the molecules on the surface from achieving an equalization in temperature with the molecules at the bottom.

Milk

How Do We Keep Milk from Boiling Over?

One of the major difficulties in science is accurately assessing situations. To what degree can one simplify a system without losing the essence of the phenomenon one wants to understand? To explain how soup is cooled, I compared it to water, because the heat exchanges on the surface, which are solely responsible for the cooling, are identical for soup and water. If we were interested in the properties of flow, only a very watery soup could be compared to water. Like the poet, the physicist and the chemist must be masters of metaphor.

To understand why milk boils over, can we also consider it analogous to water? Certainly not, because boiling water does not boil over. Clearly, milk is a more complex liquid than water. A bit of observation will reveal to us something of its hidden nature: when we let milk stand, what rises to the surface is cream, that is, a fatty substance (since cream, beaten, becomes butter). In what form is cream found in milk? Observing milk under the microscope would reveal to us countless tiny fat droplets (globules), dispersed in a solution. Milk is an emulsion, and by reflecting the light at their surface and also dispersing it, the globules of fatty material dispersed in the water are responsible for the milk's white color.

Naturally, milk is not only water and fat, because the two materials do not mix. Melted butter and water remain separated (in science, we say that they form two phases); with unmelted butter, the separation is more like divorce. In fact, milk also contains proteins and various other surface-active molecules, that is, molecules with a part that is soluble in water and a part that is soluble in fat. When the water-soluble part comes in contact with water and the fat-soluble part comes in contact with fat, these surface-active molecules form a coating that delimits the fat globules, stabilizes them, and assures their dispersion in water. This stabilization is reinforced by the casein molecules on the surface of the globules, which provide a mutual repulsion between the globules, since these molecules are negatively charged.

Nevertheless, the repulsion produced by the casein alone is not enough to prevent the occasional coalescence of globules, that is, their fusion. In a liquid, the globules are in constant motion at various speeds. Those moving the most rapidly manage to collide with one another and fuse into larger globules. Now, the larger the globules are, the weaker the forces of repulsion become in relation to the buoyancy. Gradually, the globules enlarge and rise: cream is formed on the surface of the emulsion.

When milk is heated, the effect is even more rapid, because the globules themselves are moving more rapidly. Their collisions more often lead to their fusion and at temperatures higher than 80°C (176°F), the casein coagulates. This coagulation has two effects. The coagulated casein no longer protects the globules, and it forms a continuous layer on the milk's surface, a skin. Water vapor forms at the bottom of the saucepan, gradually gets trapped under the skin, lifts that skin because its volume is much greater than the volume of water . . . and milk boils over onto the cook with the horrible odor of rotten eggs.

Why That Odor?

The odor occurs because the proteins in milk are amino acid chains, certain links of which include sulfur atoms. At temperatures higher than 74°C (163°F), these chains are destabilized, and their sulfur atoms react with the hydrogen ions in the solution, forming hydrogen sulfide. That is the substance that possesses this . . . odor of cooked milk, to put it mildly.

Why Is Human Milk More Digestible than Cow's Milk?

Few readers of these lines will have the opportunity to taste human milk, but many of us have already done so, at the beginning of our lives. And many of us—about half—have the capacity to produce it.

Why is human milk more digestible than cow's milk? Because it contains fewer proteins. Proteins coagulate in the acid environment of the stomach, so they are less available to the digestive enzymes. Thus digestion is slowed down.

The effect is the same as the one used in making cheese. Milk coagulates when salt or an acid like vinegar or lemon juice is added to it because the positively charged ions of the salt, attracted by the negatively charged ions of the casein, position themselves around them and neutralize the forces of repulsion between the globules, which can thus coalesce. The property that makes cow's milk less digestible than human milk is an advantage in the case of cheese. With more proteins, coagulation is easier. To each stage of life its pleasures. . . .

Gels, Jellies, Aspics

The Principle of the Calf's Hoof

Gels, jellies, aspics . . . At these words, the gourmand is overcome with visions of brilliant dishes where the transparency of a glossy coating haloes immense fish surrounded by herbs or slivers of truffles adorning roast fowl. Beauty is not the only characteristic of aspics. In the mouth, they melt deliciously, leaving behind the quintessence of the flavors sealed within.

The gourmand secret in these dishes is in the jelly, whose mysteries have now been revealed to us by physics, prompted by the photography industry.[1] After centuries of empiricism, today's cooks are now equipped with knowledge of the reasons behind their enchanting gastronomic creations.

For a long time, cooks have known that simmering certain meats, such as a calf's hooves, releases into the cooking liquid "principles" that at room temperature congeal the dishes in which they are present. This is how a simple chicken in aspic is prepared: the fowl is heated for a long time in a liquid, then the preparation is cooled. The solution remains transparent but jelled. That is also the secret of meat glazes, used to prepare sauces. In water, a mixture of carrots, onions, and various other aromatic ingredients is cooked with a few ground bones on which a bit of meat remains (fish bones are used for a fish-based stock). After a long simmering, collagen (the principal protein in skin,

---

1   It is a gelatin gel, spread over a plastic film, that contains the specks of silver that record the passage of light.

tendons, cartilage, bones, and connective tissue) is gradually extracted, transformed, and concentrated into a viscous syrup that can later be used to thicken a gravy or wine sauce.[2]

This recipe is simplified if, in place of the meat glaze, you use packaged gelatin, which contains almost the same collagen molecules as the ones that would slowly be extracted during the preparation of the stock. The meat glaze provides more flavor, however, because during the slow cooking process, some collagen is hydrolized, that is, dissociated into its component amino acids. But if the accompanying wine is good . . .

Different jellies are the ones prepared with fruits and sugar. Many fruits, such as apples, contain jelling agents that transform the liquid juice and sugar into a substance appreciated by children and skirted and bearded gastronomes alike. Their secret will be revealed in the chapter on jams.

Still other jellies, though less transparent, are seafood mousse, filled quiches, and even soufflés and meringues. We will examine those in the corresponding chapters. And finally let us not forget the starch gels, prepared by mixing flour or another starch and water. Their importance is so great that I will assign their examination to two chapters, one on sauces and one on pastry.

A Trap for Water

The gelatin gels that I will examine now consist almost entirely of water (with flavor, of course!). With a little store-bought gelatin, about two grams (.07 ounces), many deciliters (1 deciliter equals about .42 cup) of water can be jelled. This jelling is reversible. When heated, such gels liquefy, even if they are subsequently cooled and regain their semisolid consistency. Like jams, gelatin gels are physical gels, different from chemical gels like those of eggs. In cooking, an egg white coagulates and forms a permanent gel.[3]

---

2 This kind of meat glaze can be used, for example, to make *tournedos au Pinot noir d'Alsace*: Fry a small tenderloin steak over a hot flame in good butter, and when it is cooked place it in a hot oven. Deglaze by pouring into the frying pan two deciliters (about three-quarters of a cup) of Pinot Noir and two spoonfuls of meat glaze. Reduce it, add two spoonfuls of cream, and top the steak with the syrupy sauce that forms.

3 Of course, if you are a chemist, you can "uncook" these chemical gels!

Gels began to lose their mystery after 1920, when the physical chemist Hermann Staudinger (1881–1965) created the concept of macromolecules, that is, very long molecules, analogous to threads, sometimes (as in the case of proteins) capable of winding themselves into balls or unwinding, according to their composition and the environment in which they find themselves. Thus it was understood that macromolecules, like those of gum, gelatin, and cellulose, could be linked in an aqueous solution to form a continuous network extending through the whole mass of the solution. Only a very few macromolecules linked in this way are enough to immobilize a large amount of water, thanks to their numerous hydrophilic sites.[4] In the case of gelatin, for example, a transparent, homogeneous gel forms when the temperature of a solution is lowered to below about 35°C (95°F).

Why does gelatin form soft gels even though collagen is rigid? Because the collagen protein in animal tissues forms a fibrous structure. The solidity of the collagen fibers, which are responsible for the toughness of meat that is cooked a long time, like neck meat, is apparently due to the particular composition of the chains. Like all protein molecules, the collagen molecule is a long chain, the links of which are the amino acids (twenty different amino acids appear in animal or vegetable proteins). More precisely, the collagen sequence would be glycine, some other amino acid, proline, glycine, some other amino acid, hydroxyproline, etc.

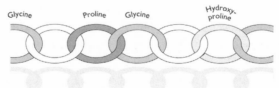

Proline and hydroxyproline are chemical groups that rigidify the chain locally, and their lateral atoms play an important role in the solubilization of the protein. They interact with the water molecules through weak bonds, hydrogen bonds.

---

4   This pedantic word should not frighten anyone. It only means that these groups bond to water molecules. It comes from the Greek *hydro-*, water, and *phile-*, that which loves. Conversely, the hydrophobic groups, from the Greek *phobos*, that which has fear, are those that do not bond to water molecules.

In animal tissue, collagen molecules spontaneously form triple helixes, stabilized by hydrogen bonds between the adjacent lateral groups of amino acids. And these triple helixes align themselves into fibers by bonding at their extremities. The triple helixes are grouped together into solid fibrils composed of about one thousand units. Structured in this way, the collagen is insoluble in cold water.

Collagen is extracted, however, by heating it in the presence of water. The water molecules insinuate themselves between the collagen molecules, separating them and making them pass into solution. The same result is obtained with the help of bases or acids. With acids, meat becomes tender because the collagen that normally toughens it passes into solution. That is part of the secret of marinades.

Why Do Gels Have to Be Made Slowly?

In gel making, the trick is to get water molecules to dissociate the collagenic tissues, separating the three units of the collagen's triple helixes into long isolated strands that tend to recombine later, when stocks cool. In fact, if a gelatin solution is allowed to cool, the triple helixes do recombine, as the appropriate sites come face to face because of the random movement of the molecules. You can observe an analogous phenomenon if you toss some loose magnets into a bag. The combinations occur spontaneously in both cases. Quickly, however, recombination flaws block this process at that stage in which a continuous network permeates the whole solution. And that, considered as a whole, is a gel, a jelly.

The important thing that physics teaches us is that the state of equilibrium is only achieved after several days, under the usual conditions for creating gels. When three strands recombine locally into a segment of a triple helix and when their combination has been blocked by a flaw, it becomes difficult for them to disengage to recombine more perfectly if the temperature is too low, since molecular motion has been slowed down. Similarly, a marble dropped near the summit of a landscape cut by valleys is not always immobilized at the depths of the lowest depression. Sometimes it falls into a small depression midslope and stops there because it does not have enough energy to continue down to a lower level.

That is why a gel prepared by slow cooling (left on a work table in a cool room, for example) is ultimately firmer than a gel that has been cooled rapidly in the refrigerator. The flaws have not been fixed in an indestructible configuration. A warmer temperature maintained for a longer time gives the helixes a chance to disentangle themselves when they become blocked and then combine more perfectly.

We can also understand why it is necessary to avoid moving the container in which the gel is forming. Over the course of the jelling process, just before the moment when the gel forms, the masses are big and very weakly connected, which makes them fragile. If the container is moved, they come apart, and the recombination process must start over, almost from scratch. The mechanisms for this rupture are not well understood. Perhaps the helixes untwist themselves; perhaps the chains break; perhaps these mechanisms take place simultaneously. In any case, the jelling process is slowed down.

Mayonnaise

Mix Oil and Water?

You take a bowl, and you pour in oil, then water: two phases separate from each other, the water, which is denser, below; the oil, which is less dense, on top. You whisk it: a few drops of water enter the oil, a few drops of oil go into the water, but as soon as the agitation stops the oil droplets rise again and the water droplets descend. The two phases separate once more.

What miracle allows the water in an egg yolk (about half the yolk and about 90 percent of the vinegar) and oil to remain mixed in mayonnaise? The secret of the preparation is in the egg yolk. Need I say that I will not linger here over the various explanations cookbooks give for why mayonnaise gets ruined? Cookbooks contain much useful information, but they also contain many mistakes arising out of the nonscientific, empirical development of the art.

First of all, let us see why oil and water do not mix. Water molecules, composed of an oxygen atom bound simultaneously to two hydrogen atoms, are linked by what are called hydrogen bonds, between an oxygen atom of one water molecule and a hydrogen atom of a neighboring water molecule.

On the other hand, oil molecules, or lipids, are snobs that do not associate with water. In ordinary oils, these molecules are triglycerides, that is, molecules in the form of combs with three teeth, composed mainly of carbon and hydrogen atoms.

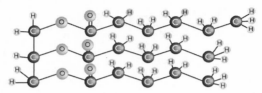

How to mix oil and water? By adding some go-between molecules to them that have an affinity for both oil and water. It is thanks to these surface-active molecules that we get mayonnaise, in which the oil concentration reaches 95 percent. Mayonnaise is an emulsion.[1] The surfactants of the egg yolk (such as proteins and lecithins) serve to coat the oil droplets, using their hydrophobic parts to make contact with the oil, and to disperse these coated droplets into the water, using their hydrophilic parts to bond to the water molecules.

Why don't the coated droplets merge into a single phase? Because the hydrophilic parts of the surfactants are electrically charged. The droplets, all presenting the same electrical charge, repel one another. This characteristic explains why acids, such as vinegar or lemon juice, help to stabilize mayonnaise. In an acid environment, certain surface-active molecules have a greater electrical charge and thus repel one another even more.

Why Does Mayonnaise Become Viscous When It Contains So Much Oil?

The more the mayonnaise is beaten as the oil is added, the more numerous and smaller the oil droplets become. Thus they occupy nearly all the available aqueous solution, flowing and moving about with difficulty. The viscosity increases.

1   The word was introduced in 1560 by the French surgeon Ambroise Paré (1509–1590), physician to three successive French kings. It comes from the Latin *emulgere*, to draw milk, because Paré noticed that many products combing oil and water are white and thick, like milk or cream.

Why Does a Dash of Lemon Juice or
Vinegar Liquefy Mayonnaise?

Lemon juice or vinegar adds water to the already constituted emulsion. The droplets have more space into which to flow. The mayonnaise is more fluid. Simultaneously, it turns whiter. Perhaps the droplets are dispersing the light differently, resulting in this effect, but that remains to be proven.

How Much Mayonnaise Can Be Prepared
with a Single Egg Yolk?

The amount of mayonnaise that can be made with a single egg yolk depends on the quantity of water present. Traditional recipes generally indicate that if there is too much oil for the quantity of yolk used, the sauce decomposes. They recommend using, at the most, one to two deciliters (3.38 to 6.76 ounces) of oil per egg yolk.

Nevertheless, my American friend Harold McGee, author of the very popular book *On Food and Cooking* (Scribner and Sons), has prepared up to twenty-four liters (25.37 quarts) of mayonnaise with a single egg yolk. Naturally, he had the aid of science. Knowing that oil arranges itself into droplets in a continuous phase of water, he figured that the small quantity of water normally contributed by the egg yolk (about a half teaspoon per yolk) was not enough to prepare a large emulsion. Thus, to maintain the separated oil droplets in the aqueous phase, he added water as he added oil. More precisely, for each cup of oil, he advises adding two or three teaspoons of water.[2]

Since a large egg yolk contains enough surface-active molecules to emulsify many quarts of mayonnaise and since too much egg yolk gives the mayonnaise a taste of raw egg that some find disagreeable, I suggest that, when you wish to prepare a small quantity of mayonnaise, you do not use the whole yolk—a drop is enough to make a big bowl of mayonnaise—and you begin the sauce with lemon, vinegar, or plain water, adding a few finely chopped herbs for flavor.

Why Must Mayonnaise Be Beaten Vigorously?

It is necessary to break up the oil into little droplets and make them migrate in the water, carrying the surfactants.[3] Now, the lower the temperature, the greater the difference between the miscibility of the water and oil. If you congeal the oil by cooling it too much, you will no longer be able to divide it into droplets. For the same reason, you must warm the butter used in preparing a béarnaise or a hollandaise sauce, two other emulsions in which the egg, again, provides the surfactants.

Why Can't the Oil Be Poured In All at Once?

Classically, recipes indicate that the vinegar (please, no mustard, otherwise your mayonnaise is no longer a mayonnaise: it is a "remoulade") must be mixed first, then the egg yolk, and finally the oil must be added, slowly, while whisking vigorously. Why add the oil to the aqueous phase, rather than the other

---

2 Twenty-four liters no longer appears to be the upper limit, as we now know that lecithins are not the principal surfactants in mayonnaise. As the French biochemist Marc Anton demonstrated at the INRA research center in Nantes, the egg proteins act more efficiently. That is why you can make much more than twenty-four liters with a single egg yolk and even more if you use the whole egg. I invite you to try the simple experiment of making an emulsion by whipping oil into egg white. The results will have no flavor, as egg whites are very bland, but you will see why using a whole egg instead of just the yolk can work when making mayonnaise.

3 It is interesting to note that some of the so-called light products owe their viscosity to tiny droplets formed by special machines used in the food industry. Why should cooks avoid such tools if they are the key to making lighter mayonnaise?

way around? First, because it is necessary to separate the oil into microscopic droplets, which is much easier if one begins with a drop of oil in water, rather than vice versa. Second, because the surface-active molecules coat the drops of oil most quickly and consistently if the surfactant is initially present in large proportions (it is initially present in the form of micelles, spheres at the center of which all the hydrophobic tails of surface-active molecules have gathered).

The first task is to produce small, well-separated drops. As long as there is more water than oil, large drops can escape the action of the whisk, and oil rises to the surface. When the volume of incorporated oil is equal to the initial volume of water and seasonings, the drops mutually prevent one another from rising, and the emulsion begins to stabilize. Then, as one continues to add oil, the small drops serve to break up the big ones, impeding their flow.

Why Does Mayonnaise Curdle?

Mayonnaise turns because it flocculates: the oil droplets merge with one another and separate from the aqueous phase. Generally, this catastrophe takes place either because the ingredients are too cold or because the emulsion does not contain enough water for the quantity of oil added.

To salvage mayonnaise that has curdled, cookbooks recommend adding another egg yolk, as if the problem were caused by the oil. But it is sometimes enough to add water and beat vigorously. You'll save an egg, but you'll need elbow grease. A better solution is to wait until the oil and water separate. Pour off the oil, and then add it back drop by drop, whipping continuously: all the useful molecules were present but not in the right configuration; you only have to rearrange them.

The Egg's Incarnations

Essential Accessories

The egg is the unrecognized star of cooking. In his *Almanach des gourmands*, Grimod de la Reynière celebrated it in these terms: "The egg is to cooking as the articles are to speech, that is to say, such an indispensable necessity that the most skillful cook will renounce his art if he is forbidden to use them."

How true! Its whites, beaten into stiff peaks, merit their own chapter. Soufflés, which they cause to rise, require the examination of so many principles of physics that, again, a complete chapter will be necessary if we are to master them. And the hard-boiled egg, though its preparation seems within the range of the least skilled novice, requires much care to be truly good.

Nevertheless, the importance of eggs in cooking is often underestimated. First of all, the egg is indispensable anytime you want to give a dish a specific form. You break an egg, whole or not, into a container that is then heated. The egg, possibly with some filling, takes the form of the container and retains it after being cooked.

Second, when its whites are beaten into stiff peaks, the egg provides the foamy element in recipes for meringues and soufflés, in mousses that are cooked, and also in recipes for the various chocolate or Grand Marnier mousses that are served cold and not cooked.

Next, eggs can form permanent gels that trap solid elements, as in, for example, clafoutis (a type of fruit tart) or quiche.

Finally, the egg is used for its surface-active compounds in various sauces: mayonnaise, béarnaise, hollandaise, gravy, and so on.[1] In all these uses, the egg is an accessory . . . an essential accessory.

In other dishes, the egg is not just an accessory but a principal player: think of boiled eggs, omelets, and eggs mimosa, for example.

Why is it so versatile? First of all, the yolk is about half water, one-third lipids (lecithin and cholesterol among them), and 15 percent proteins. The white, on the other hand, is nearly all water, since it contains only 10 percent proteins (primarily ovalbumin and conalbumin).

How does knowledge of this composition serve us? It lets us answer all the following questions.

How Can You Tell a Raw Egg from a Cooked Egg?

In a refrigerator shared by an entire family, cooked eggs are frequently mixed up with raw ones. They have the same mass (weigh them to convince yourself of this), the same color, the same surface appearance. How to distinguish them?

When in doubt, remember that a raw egg is a viscous liquid. If you make it spin, you turn only the shell. The inside of the egg remains semi-immobile, exactly like coffee when one turns the cup. Because of the friction between the liquid and the shell, a raw egg quickly loses speed, while inside the liquid slowly begins to move. A raw egg spins with difficulty and then, released, slows down. On the other hand, a hard-cooked egg, all of a piece like a spinning top, turns easily and for a long time as soon as it is set in motion. If you have no egg available for comparison, spin your mystery egg and then stop it by just touching and releasing it. A cooked egg will remain still. A raw egg will continue to spin when released because of the motion of the egg white within the shell.

---

1   See the previous chapter.

Why Does an Egg Cook?

Let us consider the simple case of the fried egg. A priori, cooking an egg is a complex operation. Think about it: all those different molecules! Nevertheless, an examination of the egg's composition shows us that what we have here, at a first approximation, is only a mixture of proteins and water.

The water behaves as expected. When it is heated, its temperature increases steadily until, at 100°C (212°F), it boils, forming bubbles.

On the other hand, the proteins are molecules analogous to long strings, often folded back on themselves because of forces that come into play between the atoms of a single molecule. When they are heated, these weak forces are broken, and since each broken bond leaves two atoms hard-pressed for companions, the heating encourages encounters between the forsaken ones, which can thus form bonds even if they do not belong to the same molecule. Moreover, some particular parts of proteins, made of one sulfur atom and one hydrogen atom, can link when the proteins are denatured. They make specific bonds called disulfide bridges responsible for coagulation.

Thus, when an egg's temperature increases, the balls of string that are the proteins begin to form chains without unwinding significantly. The liquid turns solid, but the various kinds of proteins do not all coagulate at the same temperature. One forms a network at 61°C (141°F), another at 70°C (158°F), and so on. For each temperature, there is a specific culinary result, and the higher the temperature reached, the harder the egg, because the greater the number of protein networks that trap water molecules. In the end, when all proteins are coagulated and the water is lost, the egg white becomes rubbery.

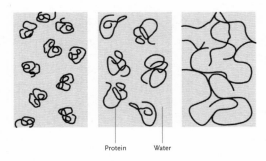

Protein   Water

Moral: When you are frying an egg, stop cooking it as soon as it turns opaque. Beyond that point, your egg will no longer be worth its salt.

Why Does the Egg Yolk Cook More Slowly than the White?

Cooks know that the yolk of an egg, fried or soft-boiled, cooks much more slowly than the white. This is partly because the major proteins in the yolks coagulate at a temperature seven degrees higher than that at which those in the whites coagulate. To complicate matters further, when boiling eggs, the white protects the yolk, causing its temperature to rise more slowly.

The famous three-minute recommendation for cooking soft-boiled eggs corresponds to the time during which the temperature increases in the various parts of an egg immersed in boiling water. After three minutes, the outside reaches 100°C (212°F) and the core reaches about 70°C (158°F), depending on the size of the egg. It takes a minute longer for the temperature of the egg yolk to rise the seven degrees necessary for its coagulation.

We must now ask, why not bake eggs in a 65°C (149°F) oven for an extended period of time (slightly more than an hour for a 60-gram egg [just over two ounces]), rather than remaining at the mercy of an egg timer? We would be sure of a perfectly cooked egg white and a perfectly runny yolk, with no risk of failure!

Why Won't the Egg White Closest to the Yolk Cook?

Anyone who has fried an egg has encountered this phenomenon: surrounding the yolk, part of the egg white refuses to coagulate.

That is because the protein in the egg white called ovomucin coagulates with more difficulty than the other proteins; this is what gives the egg white in contact with the yolk its viscosity.

How to get it to cook without letting the rest of the egg white become rubbery (see the question "Why Does an Egg Cook?")?

Salt and acids (vinegar, lemon juice, etc.) promote the cooking of a solution of proteins in water because their electrically charged atoms, or ions, come to surround the atoms possessing the complementary electrical charge in the proteins. These similar electrical charges are normally responsible for the winding and dispersing of the proteins. In the presence of complementary ions, the proteins can unwind, come together, and form bonds more easily. In other words, the proteins cook at a lower temperature in the presence of salt or acids. When cooking a fried egg, you can obtain a homogeneous white by salting the white around the yolk.

In the extreme, you can almost cook an egg by immersing it in vinegar, without heating it. The acid's ions prompt the weak bonds to break, so that the abandoned atoms can combine with the abandoned atoms of other molecules. The egg coagulates. This explanation also answers the following question.

Why Add Vinegar to the Water When Poaching an Egg?

Adding vinegar to the cooking water when poaching an egg accelerates the co-agulation of the part of the egg that is in contact with the boiling solution. The outside part of the egg coagulates immediately, constraining the rest of the egg, which can thus form a mass without dispersing into the solution. Salt is said to do the same, but experimenting will prove vinegar's superior effectiveness.

Likewise, there is an advantage to adding a little vinegar to the water used for soft-boiling eggs. If the shell cracks, the egg white coagulates immediately, sealing the leak (another solution for avoiding cracks is to make a small pin hole in each end of the egg; in that way, the air that expands does not break the egg and escapes without causing damage).

The Odorous Mysteries of the Hard-Cooked Egg

Those who know how to cook sometimes forget this: you *can* cook a hard-boiled egg badly!

Let us turn to Madame Saint-Ange, author of a long-standing cooking column whose excellent advice was collected in book form by Larousse in 1927:

> It is a very common mistake to think that there is no risk of overcooking when it is a matter of hard-boiled eggs and that therefore it does not matter how long they remain in the boiling water after they have become hard. An overcooked hard-boiled egg is

tough; the yolk is rimmed with green, the white gives off an unpleasant odor, and the whole thing gives the impression of an egg that is not fresh.

Another mistake: putting eggs in lukewarm or even cold water, and only then bringing it to a boil. The result is a faulty distribution of the white around the yolk, not achieving an attractive roundness or, in the case of stuffed eggs, attractive white cups of a regular thickness.

The consequences of the first mistake are simple to explain. When eggs are cooked too long, the egg's proteins, which contain sulfur atoms, release a gas called dihydrogen sulfide, the infamous odor of rotten eggs. This gas reacts with iron ions present in the egg's ovotranferrin (iron-carrying) proteins and thus gives it its greenish color.

To cook hard-boiled eggs properly, immerse them in water that is already boiling, allow the water to return to a boil, and let the eggs cook for ten minutes. Then put the eggs immediately into cold water. That will make them easier to shell. And again, if you want to be really modern, bake your eggs in a well-moderated oven, set at any temperature you choose, depending on how well done you like your eggs.

The Liquids in Eggs

Omelets, quiches, and the various flans are mixtures of egg whites, yolks, and a liquid (milk, water etc.). The more abundant the additional liquid, the longer the cooking time. But the heat must be evenly distributed during cooking. If the temperature in a specific spot is too high, those dreaded lumps will form.

The remedy? A pinch of flour or starch. At a high enough temperature, the long molecules of the flour pass into solution and, for reasons still unknown, block the aggregation of the egg's proteins. We will come across this effect again with regard to sauces thickened with egg yolks. It is also what lets us succeed at sabayon, custard, and the like.

A Successful Soufflé?

A Foam Beginning with a Liquid?

How to succeed at soufflé every time? Some soufflé magicians tirelessly repeat the techniques that, one day, just out of luck, guaranteed them their success. I make no claims to helping them. But what about the rest of us, whose luck has not led us to the appropriate sleight of hand? We will obtain good results more reliably if we truly understand what a soufflé is and how its constituent parts react.

A soufflé is always a foam of egg whites with some preparation added: an herbed béchamel sauce for savory soufflés or a mixture of milk or puréed fruit and sugar for dessert soufflés. The essential ingredient is the egg white, which must be whipped until stiff and to which the preparation must be added without "breaking up the foam," so that it then rises with heating and retains its risen shape after being taken out of the oven.

Let us first examine this egg white, which we whip into stiff peaks. It is a matter of a mixture of water and proteins, into which we want to introduce air bubbles. Why does an egg white foam even though water itself will not retain air? Because egg white contains proteins (essentially ovomucin and conalbumin) that, in addition to bonding simultaneously to air and water (they are surface-active), make the egg white viscous and stabilize the air bubbles introduced.

In effect, these proteins, each with a part that bonds to water and a part that repels it, tend to position themselves at the water-air interface, that is, at the border between the air and the water. In the same way that, in a mayonnaise emulsion, the surfactants of the egg yolk coat the oil droplets and disperse them in the water, the proteins of the egg white coat the air bubbles and allow their dispersion in the water of the egg white.

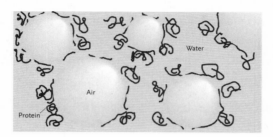

When we begin to whip the egg whites, the air bubbles are large, but the more we whip, the smaller the bubbles become.

This effect increases the stabilization of the bubbles, because the force of gravity, which normally tends to make the less dense parts of the foam (the air bubbles) rise and the liquid sink, soon becomes inferior to the force of surface tension, which is responsible for the cohesion of the air and the liquid. In other words, foam well whipped for a long time, composed of many small bubbles, is more stable than foam poorly whipped by a lazy cook.

Why does a soap solution form much less solid foam than egg whites do? Because soap molecules are generally much smaller than egg white proteins. Egg whites, more viscous than soap solutions, flow with difficulty along the interbubble surfaces. This effect is reinforced by the bonds that are established between the proteins that come to coat the surface of the bubbles. Globular proteins are long threads folded back on themselves by the forces between some of their atoms. When a solution of globular proteins is whipped, these long threads unwind, but the widowed atoms tend to reform bonds. Since they bond indiscriminately with other widowed atoms of the same protein or with widowed atoms of neighboring proteins, the neighboring proteins bond together and rigidify the water-air interface.

When Are Egg Whites Whipped Enough?

Here are a few simple formulas. For those good at calculations, first of all you will want to know that, in your vessel, an egg of 3.5 centiliters (1.18 ounces), well beaten, produces a white of 15 centiliters (5.07 ounces).

As for the more pragmatic, you will want to stop whipping when, removing your foam-filled whisk from the bowl and turning it upside down, you see that the foam remains attached to it in a solid mass taking the shape of pointed tuft, like a clown's wig.

Another trick: the egg white is whipped stiffly enough for a soufflé when it will support the weight of an egg in its shell!

In each of these cases, it is a matter of forming very small bubbles, so that the spaces between the bubbles are as narrow as possible, making it difficult for liquid to flow there, or, to put it another way, making the bubbles very stable.

Be careful! There is a danger of beating the whites too much, which separates the water from the proteins and makes the whites "weep." But this risk is not great if you beat the eggs by hand. Amateur cooks tend to whip too little rather than too much. Professionals will add sugar as a safeguard, but science still cannot explain their sleight of hand.

Why Must We Avoid Getting Yolk in the Whites?

Egg whites polluted with yolks or with fatty substances rise with more difficulty (it seems) than do pure egg whites. Why is this true? Because the yolks contain small surface-active molecules that bond to the long proteins of the whites and hamper the development of a network of the egg white proteins, weakening the water-air interface. Moreover, the fats in the yolk bond with the hydrophobic parts of the egg white proteins and thus reduce the availability of these latter to coat the air bubbles. On the other hand, when the foam is well formed, and when the egg white proteins have bonded among themselves and are properly distributed at the water-air interface, the lipids of the yolks can be added without causing harm. Places for them are no longer available.

Thus be careful with plastic bowls and, in general, containers to which fats stick. Such utensils have a harmful, if not to say disastrous, effect when whipping egg whites, because fat molecules that remain on their surfaces have the same effect as the fats in egg yolk.

What About Salt or Acid?

Many cookbooks recommend adding a little vinegar or salt to egg whites before beating them. This addition is supposed to help the foam rise and make the stiffened egg whites firmer. Is that true?

First of all, acids certainly react with whipped egg whites because their hydrogen ions (H^+) break the weak intramolecular bonds that are responsible for the proteins folding. For example, when hydrogen ions are abundant, they come in close contact with the oxygen atoms that would normally be linked to the hydrogen atoms of the same molecule. A sixteenth of a teaspoon of vinegar or lemon juice per egg white increases tenfold the concentration of the hydrogen ions, which, as small atoms of hydrogen bearing a positive electrical charge, further prevent the acid groups of the proteins from losing their hydrogen atoms and becoming electrically charged. In the presence of an acid, the proteins repel one another less. Acids also facilitate the coagulation of the proteins around the bubbles and stabilize these latter, though only to a limited extent.

Salt acts in an identical fashion, but it does not alter the dissociation of the proteins. Its ions simply come to surround the electrically charged atoms of the proteins, which weakens their electrostatic repulsion and facilitates their coagulation. Again, the effect is limited. Practically speaking, adding vinegar or salt is not of great value. It's best to stick to whipping.

A Careful Mix

What to do with egg whites after they have been beaten until stiff? If we consult them, cookbooks will suggest using them in a béchamel sauce, along with the yolks and minced, chopped, or puréed vegetables, cheese, meat, or fish. Or they may be mixed into a milk batter or with fruit purée that has been cooked and sweetened.

I would add, from experience, that mixing yolks into a basic preparation must be done away from heat, after the preparation has cooled (otherwise the yolks get cooked).

Should yolks be added two by two, as advised in certain good cookbooks? I have no idea why it should be so, but I did have a chance to test this advice in 1980, when Sunday after Sunday I inflicted on my friends the Roquefort soufflé that I was trying to perfect. I tried everything: adding all the yolks at once, adding them one by one, three by three. I obtained the best result when I added the yolks two by two. The cooks seemed right, but the mystery remained (I know now that the two-by-two trick is no use; I probably just finally learned how to make this soufflé as I was experimenting). This experience encouraged me to begin collecting and investigating similar culinary old wives' tales, proverbs, and sayings. I now have more than twenty-five thousand of them, for French cooking alone, and they are being systematically studied in the laboratory, as part of molecular gastronomy. And because molecular gastronomy is growing in popularity all over the world, I am pleased to say that people in many countries are now scrutinizing their own cuisines, collecting this old wealth of knowledge before it falls victim to modernization worldwide.

Now, back to our soufflé. Having thus mixed the yolks into the basic preparation, now comes the problem of adding the whites to the mixture. The difficulty of the operation stems from the fact that the egg whites are delicate and of a very different viscosity from that of the preparation, so they do not mix well.

Madame Saint-Ange advises pouring the lighter preparation (the beaten egg whites) over the heavier one and then cutting through both with a spatula, as if cutting a tart, bringing the heavier preparation up from the bottom and over the egg whites and repeating that operation while turning the bowl until the two preparations are thoroughly blended. Other chefs draw the heavier preparation up by beginning from the far side of the bowl and scraping along the bottom of the mixing bowl, depositing the preparation drawn from below over the top of the whites and turning the bowl with each scrape. Of course, to withstand this kind of treatment, the egg whites must be firm.

Now the mixture is poured into soufflé molds that have been buttered (so the soufflé will not stick) and floured (so that it can rise easily), being careful to fill the molds only two-thirds full (so that they only moderately overflow when the soufflés rise).

In a soufflé prepared like this, its success is due to the ovalbumin in the egg white (50 percent of the egg white), which is not denatured during the mixing and coagulates when cooked, thus limiting the expansion of the air bubbles, which would otherwise eventually explode.

Soufflés were said to rise because the air bubbles swell under the action of heating (the air expands), but a simple calculation shows that this effect can only explain a swelling of 30 percent at the most (even considering the increase in pressure, which I have not measured). If the soufflé doubles or even triples in volume, that is because the water evaporates and the resulting vapor enlarges the bubbles. Again, the coagulation of the egg proteins traps the bubbles within the mass for good.

Why Must the Oven Door Remain Closed While the Soufflé Is Baking?

Egg proteins that have not yet coagulated have not yet formed a rigid armature. Till they do, it is the soufflé bubbles, in equilibrium with the oven air, that support the weight of the preparation. If the oven door is opened before coagulation occurs, the sudden drop in temperature causes the air bubbles to contract and the vapor bubble to recondense, and the soufflé falls. Then, after the door is closed again, the walls of the bubbles coagulate before the bubbles can reinflate.

At What Temperature Must the Soufflé Bake?

The question posed in this heading requires an evasive answer. A soufflé has to bake at a temperature high enough for the proteins to coagulate before the bubbles begin to explode and the foam collapses but low enough for the interior to rise before that same coagulation prevents it from doing so. By experimenting, chefs have determined that the ideal temperature is about 200°C (392°F) for obtaining a moist center with a golden crust and about 150°C (302°F) for achieving more uniform results.

The baking time depends on the size of the soufflé. Often it is recommended that large soufflés should bake for twenty-five to thirty minutes, and small ones for fifteen minutes.

One final trick: for a nice even soufflé crust, put the soufflé under the broiler for a few moments before baking it. The top will form a solid roof, which will then rise uniformly, lifted by the air and vapor bubbles.

How Do You Avoid a Fallen Soufflé?

Ah ha! That is the great soufflé question! Some chefs advise preparing soufflés in advance, before the guests arrive, and placing them in a warm water bath until it is time to bake them. The gentle heat of the water will cause the soufflés to rise very slowly, and—a scientific mystery—they will not fall after being baked. Is this good advice?

In collaboration with Nicholas Kurti, I have studied the rising of soufflés and especially the validity of this advice. We prepared a cheese béchamel sauce, beat egg whites until stiff, and proceeded to mix them together. We then filled many small porcelain soufflé ramekins with this mixture and began by baking just one of these soufflés. Its volume had tripled, and it was perfectly baked after twenty-five minutes in a 180°C (356°F) oven. Coming out of the oven, the soufflé was beautiful, but it fell.

One of the other ramekins was placed in the refrigerator, another in the freezer, and the last two were left at room temperature. Later, the ramekin in the freezer was taken out, and when it had reached room temperature it was baked at the same time as the soufflé that had been in the refrigerator. It rose a bit better, but the result was less interesting than the first soufflé we baked. The last two soufflés were baked at the same time as well, but one of them was first placed in a warm water bath and the other one was not. They did not yield the expected results.

This series of experiments is open to criticism. The soufflé in the warm water bath had been kept in the refrigerator before this recommended treatment. Thus I repeated the experiment many times, putting the soufflé mixture in the warm water bath immediately after mixing the whites and the cheese béchamel sauce, and testing different temperatures and lengths of time for the water bath before baking. These soufflés never rose as well as the soufflés baked immediately. It is true that the cheese soufflés kept in the water bath did not fall—but that was because they did not rise.

This experiment offers the following important information: to make a beautiful soufflé, do not wait to bake it. Putting it in the freezer is only a stopgap measure, as is any time in a warm water bath.

Three Rules for a Successful Soufflé

There are three rules to keep in mind for a successful soufflé, all based on the fact that a soufflé swells primarily because of water vaporization.

First, imagine what would happen if water evaporated from the top of the soufflé preparation. No swelling would take place. Thus the highest soufflés will result from heating the ramekins from below. Put them directly on the lowest rack in a heated oven.

Second, the bubbles have to be trapped for them to be able to make the soufflé rise. Thus the egg whites must be beaten into very firm peaks, so the bubbles will be trapped in the very firm foam.

Third, the increase in volume is diminished if the bubbles escape from the top. Thus allowing a crust to form on the top of the soufflé, as proposed earlier, will promote successful rising.

To summarize, then: heat from below, use firmly beaten egg whites, and let a crust form on the top of the soufflé.

Cooking

The Secrets of Tenderness

As mentioned earlier, in cooking we have three objectives: to kill harmful microorganisms that contaminate food, to change the consistency of foodstuffs, and to give them flavor. Often the cook achieves these ends simultaneously, because heat, which kills microorganisms, for example, also degrades meat's toughest molecules and triggers chemical reactions that engender aromatic compounds.

Why does cooking tenderize meat? Why does it make vegetables lose their rigidity? And, more generally, what is the secret of culinary transformations?

The culinary arts begin and end with the art of cooking, and it certainly isn't feasible to summarize the whole book in this one chapter; such an extraction would be too concentrated. I will only attempt a preliminary clarification, saving the various types of cooking for later chapters.

In every case, heat increases the mobility of atoms and molecules, which can thus react because they now have enough energy to be transformed. The molecules bump into one another, break apart, and, when chemical groups with some affinity come into random contact because of the erratic movement of the molecules, rearrangements take place. These rearrangements are called chemical reactions. They produce new compounds.

How Do We Heat Foods?

The question seems naive, but try, for example, to heat a thick soup by holding a small electric heater over its surface. The very top layer will boil, but the rest of the soup will remain cold. Likewise, a good roast on a spit must be done in front of the fire rather than directly over it. If you roast quail, the most delicate of birds, barded as it should be with grape leaves between the fat and the bird, you will get only the taste of smoke if you cook it directly over the fire. But if you know the mechanics of radiation and put the spit at the same height as the flame and beside it, your fowl will roast gently, through to the center, and the flesh will be delicately flavored with the subtle taste of grape leaves.

This is why it is especially useful to know that heat is transmitted to dishes through three mechanisms: conduction, convection, and radiation.

BILLIARD BALLS IN THE SAUCEPAN

Conduction is the phenomenon that occurs when a solid is heated. A metal spoon left in a boiling liquid will burn your fingers. Likewise, the heat conveyed by an oven to the surface of a roast, for example, is gradually transmitted to the molecules of the interior. In effect, the outside molecules, agitated by the heat, collide with molecules located deeper within and thus transmit their energy to them. Those molecules, colliding with molecules even closer to the center, pass along the activity and thus the heat.

Heat is the agitation of molecules. Foods are like heaps of billiard balls. If you agitate the outer balls, they will transmit their agitation to the adjacent balls closer to the center of the heap, which will then transmit their energy to the balls still closer to the center, and so on. This is the phenomenon of heating through conduction.

CONVECTION AND SKIMMING

In liquids, it is convection that accelerates the transmission of heat. When a saucepan full of water is heated from the bottom, the water at the bottom heats first, through contact with the bottom of the pan, which is itself in contact with the stove. Naturally, the heat is transmitted through conduction, but beyond that, the hot water at the bottom, less dense than the cold water found above it, rises and is replaced by the cold water, which is then heated. The liquid currents, called convection currents, circulate in the liquid and rapidly distribute the heat.

This phenomenon is apparent when you skim a sauce. Say you are making a braised dish, for example. First you sear a piece of meat and its trimmings in the oven (see the chapter on braising). After five minutes of searing, you add one-fourth of a liter (about a cup) of good white wine, cover the pan, and continue the cooking at a very low boil by lowering the oven temperature to between 60°C (140°F) and 100°C (212°F), depending on the coagulation point you desire. After an hour of this slow cooking, you combine half the liquid with a roux (made by cooking a little flour in butter until it is a pleasing golden color). The roux will thicken the liquid drawn from the meat pan. Skimming it will produce a sauce that is light, smooth, and satiny; the process must be slow and consistent. First, you tilt the saucepan to one side (by raising the opposite side with an object that does not mind the heat, like an old spoon). Heated in just one spot, the liquid forms a single convection cell that absorbs the heat from below at the point of contact, rises vertically and descends along the sides. This prompts the coagulation and aggregation of all the unwanted solid matter, which is returned to the center of the convection cell and forms a scum that can be regularly drawn off. At the same time, you dab the surface of the sauce with absorbent paper toweling to remove the fat that appears as unemulsified droplets. With this slow process, you allow the time for many chemical reactions to occur . . . and to generate a wonderful flavor.

Convection was discovered by Count Rumford (Benjamin Thompson), an inspired adventurer who, among his many other scientific studies, pondered the question of why his applesauce remained hot long after his soup had cooled.[1] We know now that convection is most active when the medium is least viscous. In a viscous medium, the liquid circulates with more difficulty. In the case of Rumford's thin soup, convection rapidly exchanged the heat of the bowl and the air, whereas the heat of the thick applesauce could escape only through conduction and thus slowly. So the sauce stayed hot longer than the soup.

---

1   Count Rumford also discovered that heat and mechanical work are of the same nature. Founder of the Royal Institution in London, he met with bad times later in life because of his marriage to the widow of the father of modern chemistry, Antoine Laurent de Lavoisier.

The third method of heating involves radiant heat, which is what makes the front of you warm but leaves the rest of you cold when you are facing a fire. Heating through radiation is the principle behind roasting meat. A fire or a grill emits heat rays analogous to light rays but invisible: infrared rays. Like light, they disperse in straight lines and are stopped by opaque bodies. When they are absorbed by meat, their energy heats and cooks it.

Cooking with microwaves, of course, is also a process of heating with radiation. But in that case the waves penetrate the foods in much the same way as light passes through glass windows.

What Kind of Cooking for Which Dish?

Once the heat is in the food, it fulfills various functions that are all part of the cooking, among them, softening hard substances, coagulation, inflation or dissolution, transforming the appearance, reduction or extraction of juices or nutritional elements.

Convection, conduction, or radiation? The following analysis considers most of the cooking processes. The heating medium will generally be a fat, a liquid, dry air, or moist air.

When the heating medium is a hot solid, conduction produces a grilled dish.

When the heating medium is hot liquid fat, convection and conduction produce a sautéed dish if the food is simply lying in the fat.

When the heating medium is hot liquid fat, convection and conduction produce a deep-fried dish when the food is immersed in the fat.

When the heating medium is warm liquid fat, convection and conduction produce such wonderful dishes as confits.

When the heating medium is a boiling liquid, convection and conduction produce a boiled dish.

When the heating medium is a simmering liquid, convection and conduction produce a poached dish.

When the heating medium is steam, convection produces a steamed or braised dish.

When the heating medium is moist air, convection is responsible for the cooking of oven roasts.

When the heating medium is dry air, radiation is responsible for roasting foods on a spit.

Keeping in mind that "The discovery of a new dish does more for human happiness than the discovery of a star" (Brillat-Savarin),[2] let us not forget microwaves, which cook in a unique fashion. Radiation is absorbed by certain molecules within the food (water molecules), and the heat of these molecules then cooks the entire food by being transmitted through conduction to the molecules unaffected by microwave radiation.

Cooking Without Heat?

No list of cooking processes would be complete if it failed to mention one slightly peculiar process: cooking with chemicals.

Do not be frightened by the name. Cooking with acids, for example, is simply a matter of placing foodstuffs in lemon juice or vinegar. In these two cases, the liquid, an acid, is responsible for making the proteins coagulate. That is how fish left in lemon juice is transformed, just as if it had been poached in boiling water, although its flavor is very distinctive.

---

2 Jean Anthelme Brillat-Savarin, "Aphorisms of the Professor," aphorism 9 in *The Physiology of Taste; or, Meditations on Transcendental Gastronomy*, trans. M. F. K. Fisher (New York: George Macy Companies, 1949; reprint, New York: Counterpoint, 1999), part 1, p. 4.

The Boiled and the Bouillon

Fifty Hams!

For a long time, meat was believed to be made up of two parts: the part that passed into the bouillon when meat was cooked in water for a long time and the fibrous part, which was called *le bouilli*, or the boiled. Gourmands did not have words harsh enough for *le bouilli*. Stripped of its succulent elements, boiled meat no longer deserved a place on their tables.

Brillat-Savarin relates the following anecdote in honor of the "osmazome," which was supposed to be the principal component of taste in meats:

> The Prince of Soubise planned to give a great party one time; it was to end with a supper, and he asked for the menu.
>
> His steward appeared at his morning conference with a handsomely decorated sheet of paper, and the first notation the prince's eyes fell upon was this: *fifty hams*.
>
> "Look here, Bertrand," he said, "it seems to me you are dreaming! Fifty hams! Are you trying to treat my whole regiment?"
>
> "Not at all, sir! Only one ham will appear on the table, but the rest are essential for my *sauce espagnole*, my white sauces, my garnishes, my . . ."
>
> "Bertrand, you're thieving from me, and I shan't let you get away with it."
>
> "Ah, my lord," the artist said, hardly able to hold back his wrath, "you know very little of our resources! Command me, and I can put these fifty hams which seem to bother you into a glass bottle no bigger than your thumb!"

What was there to say to such a positive assertion? The prince smiled, nodded, and the menu was approved.[1]

A bit further on, Brillat-Savarin considers the osmazome more explicitly:

The greatest service rendered by chemistry to alimentary science is the discovery or even more, the exact comprehension of osmazome.

Osmazome is that preeminently sapid part of meat which is soluble in cold water, and which differs completely from the extractive part of the meat, which is soluble only in water that is boiling.

It is osmazome which gives all its value to good soups; it is osmazome which, as it browns, makes the savory reddish tinge in sauces and the crisp coating on roasted meat; finally it is from the osmazome that come the special tangy juices of venison and game.

This property is found mainly in mature animals with red flesh, blackish flesh, or whatever is meant by well-hung meat, the kind that is never or almost never found in lambs, suckling pigs, pullets, or even in the white meat of the largest fowls. It is for this reason that lovers of poultry have always preferred the second joint: in them the instinct for flavor came long before science confirmed it.

It is also the infallible goodness of osmazome which has caused the dismissal of so many cooks, destined as they were to ruin their basic soup stock; it is osmazome which has made the reputation of the richest consommés, which once made toast soaked in bouillon a favorite restorative during weakening curative baths, and which inspired Canon Chevrier to invent a soup pot which locked with a key. (It is this same holy Father who never used to serve spinach on a Friday unless it had been cooking since the Sunday before, and put back each day on the stove with a new lump of fresh butter.)

Finally, it is to husband this substance, as yet largely unrecognized, that the maxim has been propounded that in order to make a good bouillon the pot must only *smile* with heat, a truly worthy expression considering the country from which it came.

Osmazome, discovered at least after having for so long delighted our forebears, can be compared with alcohol, which tipsified many generations of men before any of them knew how to strip it naked in the analytical process of distillation in a laboratory.

---

1 Jean Anthelme Brillat-Savarin, "Meditation 3," sec. 20 in *The Physiology of Taste; or, Meditations on Transcendental Gastronomy*, trans. M. F. K. Fisher (New York: George Macy Companies, 1949; reprint, New York: Counterpoint, 1999), part 1, pp. 53–54.

During the action of boiling water osmazome gives place to what is understood more especially by extractive matter: this last product, reunited with the osmazome, makes up the juice of meat.[2]

A Universal Flavor?

Even if it is scholarly and learned, this long treatise by our master gastronome is very wrong. Brillat-Savarin's osmazome is only a myth created in the era when analytical chemistry was in its infancy. It was the French chemist Louis Jacques Thenard who coined the term "osmazome," based on the Greek *osmé*, "odor," and *zomos*, "soup." He proposed it for the first time in an article in the *Bulletin de la Faculté de médecine de Paris* in 1806.

In his use of the term, Brillat-Savarin seems to suggest that the osmazome is a unique, well-defined compound, like the ethyl alcohol in alcoholic drinks. But modern methods of analysis show that the part of meat extracted cold is already a complex mix of water, lipids, various odorant molecules, salts, and more. In total, meat contains hundreds of sapid or odorant compounds. As for the first extract being the most sapid, let us trust our ancestors on this. They were more accustomed to boiling meat than we are, and, what is more, volatile molecules are often better perceived than the molecules the food retains, which activate neither the taste buds nor the nasal receptors.

On the other hand, Valéry's aphorism according to which "what is simple is always false" applies here to the osmazome. It is not the sapid element in meat; it is only one of the various flavorsome extracts that can be drawn from it.

And, if we are to believe Brillat-Savarin, it is the best.

How Does One Obtain a Flavorful Bouillon?

Let us think this through: meat contains many proteins. That is well known, and we eat meat for the protein it provides us. But what else does it contain? Amino acids, as these are produced when proteins are cooked for a very long

2 Idem, "Meditation 5," in ibid., part 1, pp. 66–67.

time. These compounds are important because, as we have seen, they provide taste. What else? Fat! Just as important, but less well known, is the fact that meat fats are the main storage sites for odorant molecules. Beef tastes like beef because its fat contains the odorant molecules characteristic of beef. Mutton tastes like mutton because its fat contains the odorant molecules characteristic of mutton. And if little birds—larks, robins, buntings—each have their own delicate flavor, that is also because their fat contains the molecules characteristic of this small game.[3]

How to obtain a flavorful bouillon when the odorant compounds are in the fats, which are eliminated after cooking, once they have congealed?

First of all, let us note that compounds insoluble in water are never completely insoluble. In the presence of oil and water, they divide in specific proportions in each of the two substances. To extract a compound from oil that is not very soluble in water, one must simply agitate the oil in the presence of a large quantity of water. Then, when the extraction has been achieved, one can eliminate the oil and concentrate the water . . . and the flavors.

In practice, we can trust Brillat-Savarin when it comes to obtaining a good bouillon, even if his instructions only give the appearance of being scientific:

> The water first of all dissolves part of the osmazome; then the albumen, which coagulates at about 104 degrees Fahrenheit, forms a scum which is usually skimmed off; then the rest of the osmazome dissolves with the juice or extractive part; and so finally do portions of the outer coating of the fibers, which are pulled off by continuous movement of the boiling liquid.
>
> To make a good bouillon, the water must heat gradually, so that the albumen will not coagulate inside the meat before it can be extracted; and the boiling must be kept at a simmer, so that the various parts which are successively dissolved may mix together easily.[4]

---

3 Gourmands are well aware that meat fats contain their flavors. Especially when preparing small game birds, they wrap them in bacon, skewer them, and roast them before a fire, over slices of bread that catch the drippings and on which the liver is then crushed.

4 Brillat-Savarin, "Meditation 6," sec. 32 in *The Physiology of Taste*, part 1, p. 76.

Does modern science confirm these precepts? It teaches us that the muscle fibers (the cells that make up the muscles) are composed of two proteins essential for contraction, actin and myosin, which are coated in collagen fibers.

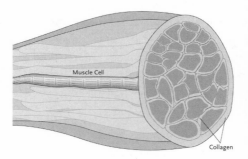

It is the collagen that rigidifies meat and makes it tough. And it is to tenderize the collagen that meat is cooked for a long time in water. Over a long period of cooking, the collagen gradually passes into the bouillon at the same time as it is partially decomposed. That is how gelatin is extracted from meat (or from bones, skin, and tendons, where gelatin is abundant).[5]

So why does the boiled beef in a stew retain its fibrous texture? Because, even if the collagen is solubilized, the proteins within the muscle fibers coagulate and are not solubilized.

Why should bouillon be covered when cooking? Because, as the bouillon boils, odorant molecules escape with the steam. As chefs will tell you, however, covering bouillon turns it cloudy. Why does it turn cloudy? That I do not know. How do you clarify a bouillon if it should become cloudy? By beating in an egg white, which coagulates around the particles that cloud your bouillon. You can also use good filtering tools from a chemistry laboratory, which are more effective and do not waste egg whites.

Finally, why don't the volatile molecules of the meat escape from the bouillon when the bouillon only simmers? That is the crux of the whole matter.

---

5 This is the same gelatin that, when present in glazes, demi-glazes, and various stocks, serves to bind the sauces without egg, roux, puréed vegetables, or blood.

First one should remember that the fat that melts during cooking remains in the stock as fat droplets, dissolving odorant molecules. Certain volatile—and odorant—molecules actually do leave the bouillon, but they react as well. They make new odorant molecules that enrich the stock, according to Brillat-Savarin, who was not entirely mistaken. Over the course of the cooking, Maillard and other browning reactions produce many sapid molecules that enrich the bouillon. The bouillon's flavor is primarily the result of this cooking process. Someday, compare the water a piece of meat has been steeped in cold, in which such browning reactions have not taken place, and a bouillon prepared according to these worthy, time-honored principles.

Can You Lose Weight by Eating Only Boiled Meat?

That is a serious question in our times, when we often do not get the physical exercise that would permit us less restraint at the table. We have seen that meat contains abundant protein but also fat, which gives it its desirable flavor. Could this flavor be retained by boiling meat?

In red meat heated to 150°C (302°F) or in white meat heated to 240°C (464°F), the fats melt and are released from the flesh. In cooking, meat gets rid of its fat. If grilled meat is sometimes criticized for being too fatty, it is precisely because it is coated with the fats that have issued from it.

Can the problem be remedied by boiling meat? Not necessarily. For a bouillon, the maximal temperature is 100°C (212°F), so the fats melt less and are released from the meat less easily. Additionally, the mineral salts and aromatic compounds of the meat cells can escape from the meat and pass into the bouillon. In sum, we obtain a less flavorful—and also less healthy—food.

It is better to wipe grilled meats with absorbent paper. This will allow you to retain the good taste (the taste produced by grilling, not the taste of the fats that you'll eliminate) while preserving your figure.

Steaming

How Does One Tenderize Without Sacrificing Flavor?

Boiled meat can be tender, certainly, but it has no flavor. In the last chapter, we saw that the tough part, the connective tissue and especially the collagen, is dissociated as it reacts with the water, but the odorant and sapid molecules escape from the meat into the bouillon. All that remains are tasteless fibers.

Can the flavor of the meat be retained even while it is tenderized? That is the principle behind cooking with steam, no different from cooking in a sealed pot (*à l'étouffée*), a long cooking process in an atmosphere saturated with water vapor. During this operation, the saucepan acts as a papillote, a sealed tinfoil or parchment packet. The same principle is at work in a Texan barbecue, when the meat, set on a grill in a big canister full of coals, cooks for sometimes as long as two days at just 70°C (158°F).

Thus, for steaming, it is sufficient to have the food steep for a long time in vapor. Naturally, the hotter this vapor is, the more rapidly the food cooks. That being the case, you need a lively boil. Of course, the food must be above the liquid, or else you end up with boiled meat!

Often recipes recommend browning meat in butter before adding the liquid and a little salt. This is not a bad idea. First, the tenderizing can get started in this first stage of cooking at a high temperature, and, what is more important, it promotes Maillard reactions and the browning that produces the characteristic odor of grilled meat.

After the preliminary browning, liquid is added, above which the meat is raised—in the basket of a pressure cooker, for example, not used under pressure—and cooked for a long period—four to five hours—during which time the collagen tissue is dissolved. This method is especially suited to dishes in which flavors like those in a bouquet garni are to be highlighted. The herbs, spices, and seasonings are added to the liquid so that their aromatic compounds can be extracted by the water vapor (extraction by vapor is a method for separating compounds widely used in chemistry laboratories and also in the perfume industry), conducted around the meat, and recycled. Brought to a temperature never higher than 100°C (212°F), they are not degraded and gradually permeate the surface of the meat. Let us not overlook a method that offers such advantages!

Braising

Meat for Cooking Meat?

I would love to introduce you to the notion that the supreme cooking method is braising. In this transformative operation that takes place in a closed receptacle, with almost no liquid, the meat loses as few elements as possible. Instead of the elements of the meat escaping into the liquid, the meat absorbs the best of the liquid.

Before going further, let us recall cooking's old familiar refrain: to kill microorganisms, provide flavor, and make tender. In braising, these operations take place in two stages: cooking at a high temperature, which kills microorganisms, browns the meat's surface, and creates odorant and sapid molecules through Maillard reactions; then a very long phase of tenderization and taste production using gentle heat. The result measures up to the effort: the meat, marvelously flavorful, melts in the mouth.

People often believe that braising requires immense care, and this fear often makes them substitute roasting for braising. There is no need for this. If one sets about it methodically, braising requires no more attention than roasting, and its success is certainly more guaranteed. What is true, nevertheless, is that braising comes at some expense: braised meat cooks in the sapid juices of some other meat.

The principle of braising, then, is to cook the meat while at the same time nourishing it with fortified juices. This is a long way from cooking in water! It is not only the method of cooking that characterizes braising but, just as important, what is added to the pot: bacon, meat juices, wine, brandy, all of which give succulence to the piece of meat being cooked in it. Succulence is the true goal of braising, the one toward which all the operations and various methods are aimed. According to the great Carême, to braise is "to put strips of bacon in the bottom of a casserole and, on top, slices of meat. Then one adds either a goose, a turkey, a leg of lamb, a piece of beef, or something similar. Then one adds slices of meat and strips of bacon, two carrots cut in pieces, six medium onions, bouquet garni, basil, mace, coarsely ground white peppercorns, a touch of garlic; then half a glass of well-aged brandy and two large spoonfuls of consommé or bouillon. Then it is covered with strong paper that has been buttered, and cooked from above and below."[1]

These instructions are not presented as a recipe to follow step by step; rather, they give us an idea of the philosophy of braising. Let us consider a recipe: in a large pot that will be closed and go in the oven, one puts a fatty substance (oil or goose fat, for example), a layer of onions, a layer of carrots cut into rounds, ham, the piece of beef that will be eaten, strips of bacon, another layer of ham, another layer of carrots, and a final layer of onions. Without covering it, the whole thing is placed in a very hot oven to brown. Then bouillon, meat juice, white wine, and possibly brandy are added. Then the covered dish is put back in the oven, at very gentle heat, so that the braising is indicated by only a very slight trembling. At the end of the cooking, the remaining juices are recovered and bound with a light roux. Then the sauce is skimmed while the meat is kept hot.

---

1   Marie-Antoine, or Antonin, Carême was born to a very poor family in Paris in 1783. Abandoned to the streets at the age of ten, he had the luck to be taken in by the owner of a cheap restaurant who taught him to cook. Struck by his talent and his desire to learn, the pastry chef who was his master when he was sixteen helped him with his studies and secured him access to the collection of engravings at the National Library, where he copied architectural models, which, replicated in pastry form, were admired by the premier consul. After directing the kitchens of Talleyrand, England's prince regent, Czar Alexander I, the Viennese court, the English embassy, Princess Bagration, the lord steward, and the baron de Rothschild, he died at the age of fifty, "burned, so they say, by the flame of his genius and the coals of the rotisseries." This "Larmartine of cooking" published surveys on the cooking of his day, making himself one of the pioneers in so-called monumental cuisine, of which Urbain Dubois is the chief representative.

What Happens During Braising?

The principle behind braising, I repeat, is to tenderize the meat and make a very flavorful sauce. Here I must cite some new scientific results, which disprove some old theories. Previously, it was thought that there were two main kinds of cooking: by expansion and by concentration. In the first case, however, exemplified by boiling meat, no expansion of the meat takes place. On the contrary, it shrinks, because the collagen contracts when it is heated and the juices that flavor the stock leave the meat. In roasting or in braising, on the other hand, there is no concentration, as it was thought, but again some contraction because the collagenic tissue shrinks.

It is better simply to remember that, the higher the cooking temperature, the greater the loss of juices. In braising, the low temperature that should be the rule keeps as many juices in the meat as possible, while the collagenic tissue dissolves slowly, releasing gelatin and amino acids that give the sauce taste and a satiny texture.

Braised Meat Without Sauce?

Braised meat without sauce? That would be a crime against gourmandise. In fact, the sauce for braised meat is not difficult to prepare: either the juices thicken enough naturally to be served as they are or binding them with a bit of potato flour, roux, or beurre manié completes the preparation of this delicious, mouthwatering dish.

Let us remember that a roux is prepared by cooking flour very slowly in butter. The mixture of butter and flour must form bubbles that rise gently and then collapse again. The roux is ready for the liquid it will bind when it has a nice hazel or light brown color, according to the taste desired. After the juices have been added to the roux, skimming completes the sauce's preparation. The thickened sauce is heated for a very long time, keeping only one corner of the saucepan in contact with the heat, so that a single convection cell agitates the liquid. The top of this cell is skimmed to eliminate all solid particles that cloud the sauce. By eliminating excess flour and fat particles, this operation has the huge advantage of producing a healthy product that is, at the same time, flavorful.[2]

Beurre manié, a raw butter-and-flour mixture, is used when there is not time enough to make a roux. Butter and flour are mixed with a fork, the liquid to be thickened is brought to a boil, and balls of the blended butter and flour are added to it.

---

2 Indeed, remarkably flavorful, as we demonstrated in the laboratory when we tested sauces made only from pure water, pure starch, and pure fat. After about two hours of skimming, some mushroom flavor was produced! I do not yet understand why this happened, but it certainly indicates that the old recommendation to cook sauces for a very long time was sound. Chemistry slowly improved the flavor of the sauce; after the first clarification, skimming probably only served to prevent the formation of a skin on the surface.

Chicken Stew, Beef Stew, Veal Stew

How Do We Salt Them?

When should salt be added to a beef stew, a veal stew, or one of those chicken stews of good King Henry?

Try this if you have not already considered this question. One day when you have a little more time than usual at your disposal, make double the amount of a dish and experiment with the effect of salt. In two saucepans heated in identical fashion, place the same ingredients in equal quantities, but salt one of the dishes before cooking and the other one after. You will soon see the difference . . . and the importance of osmosis. It is a matter of simple physics, as a familiar experiment will reveal. When a drop of colored liquid is added to pure water, and they are left together for a bit of time, we discover that the coloring agent eventually spreads throughout all the liquid.

The molecules of the coloring agent, agitated by incessant movement and randomly bumping into the water molecules, spread throughout the water, and their concentration becomes equalized throughout the solution. This phenomenon of diffusion is very common. In a medium where molecular movement is possible, compounds gradually distribute themselves so that their concentration is everywhere equal.

Let us complicate the experiment a little by dividing a U-shaped tube in half with a permeable membrane that only lets water pass through and stops any larger molecules, like those of the coloring agent, putting water in one compartment and the coloring agent in the other. In order to distribute itself equally throughout, the water will go into the compartment containing the color to equalize its concentration; the color molecules, however, will remain in their initial compartment because they will be stopped by the membrane.[1] In the end, the compartment that first contained only the coloring agent will gain a bit of water, so the levels will be different. This is the phenomenon of osmosis.

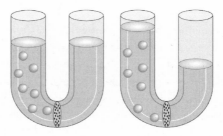

The dilemma of when to add salt to the stew is as follows: if the dish, basically a piece of meat in water, is not salted at the start, the mineral salts pass into the solution in which the meat steeps. At the end of the cooking time, the meat is tasteless. On the other hand, if it is salted before cooking, the gravy will suffer because the meat's juices will remain within the meat.

You must thus proceed according to the spirit of the dish that you are preparing. If you want to enrich the bouillon, add salt only at the end of cooking. If you want to retain the full flavor of the meat, add salt right away. And if you want a good chicken stew, with a flavorful sauce, add salt a little before the end, so the juices are harmoniously divided between the two components of the dish.

---

1 My more rigid colleagues will be angry to read this, because molecules have no will. Of course, I employ this way of describing physical phenomena simply as an aid to comprehension.

Questions of Pressure

Why Use a Pressure Cooker?

The pressure cooker is an antimountain. At higher altitudes, the air becomes rarified, and the air pressure is lower than that at sea level, so water molecules, for example, more easily escape the body of liquid in which they are found. In short, water boils at a temperature lower than 100°C (212°F). In a pressure cooker, the water that evaporates at the beginning of the cooking gradually increases the pressure in the pot, so water molecules have more difficulty escaping the liquid. The boiling point of the water is thus increased. In practice, today's pressure cookers are devised so that water boils in them at 110°C (230°F) to 130°C (266°F).

This increase in the boiling temperature has some advantages. Chemical reactions take place about three times faster in water at 130°C (266°F) than in water boiling at 100°C (212°F). Vegetables, for example, cook much more quickly.

On the other hand, the pressure cooker has some disadvantages that make cooks condemn it. First of all, you cannot see what is happening inside it and controlling the cooking process is more difficult. Five minutes too long in the pressure cooker is like fifteen minutes of traditional cooking. In addition, certain reactions that take place in an uncovered saucepan, involving the air in the kitchen, do not occur in a hermetically sealed pressure cooker.

Moreover, all reactions are not accelerated in the same way by the increase in temperature. The softening of vegetable fibers is more accelerated than the permeabilization of the vegetable cell walls. The vegetables are tenderized, but they remain tasteless. Some cooks will only pressure-cook a roast as a last resort. For others, it is a method of choice; they maintain that a pressure-cooked roast is less dry than an oven roast.

Let us not sit in judgment of the pressure cooker here. Let us only try to understand the principle behind it.

Cooking in the Mountains?

Since we have considered cooking under increased pressure, why not also consider cooking under decreased pressure?

Those of you who suffer from vertigo may rest easy. I am not inviting you into high altitudes. I am only proposing a simple device, the vacuum pump, which reduces the pressure in a receptacle to which it is attached. Present in all chemistry laboratories, the vacuum pump is a simple tube that can be connected to a faucet to allow water to run slowly (great for watering your culinary herbs, for example). This tube includes a lateral branch where a plastic tube is fitted, the other end of which can be attached to the opening of a pressure cooker where the safety valve is usually located. As it flows, the water takes in the air and creates a partial vacuum in the pressure cooker. Because of the phenomenon mentioned at the beginning of this chapter, the boiling temperature is lowered. This time, the chemical reactions are slowed down in various ways; new tastes appear.

We tested this ingenious assembly of Nicholas Kurti's during the First International Conference on Molecular and Physical Gastronomy, held in Erice, Sicily, in 1992, but the results remain unexamined. We know that a bouillon reduced under low pressure has a different taste, but the circumstances that call for such tastes have yet to be discovered.

Cooks, the ball is in your court!

Roasting

First Principle: Succulence

A quick cooking process that is not meant to tenderize the meat, roasting is reserved for choice pieces that come from young, tender animals. It leaves the characteristic flavor of the meat intact, adding only a delicate touch on the surface. The juices, most of which remain in the flesh, flood the mouth with subtle flavors when one bites into the meat.

A roast retains its succulence only when it is seared. The oven must be preheated, the meat must be coated with oil, which conducts heat better than water, and ideally steam must be eliminated. Without allowing too many of the juices to be released, the heat produces new sapid and odorant compounds by destroying various molecules at the surface and thus blending the contents of various cell compartments, which react with one another. The lipids (fats), amino acids, and sugars bind together into large, dark-colored odorant and sapid molecules.

Traditional French cooking requires beef, for example, to be roasted in such a way that the center remains rare. The surface is covered with oil or butter so that, seared more rapidly, it forms the thin crust so prized by gourmands.

A long time ago, it was believed that searing produced an impermeable layer that would keep the juices from escaping. But it has been demonstrated that the juices escape regardless. If meat should be roasted in a hot oven, it is to limit the roasting time and thus the time during which the juices can escape. When the roast comes out of the oven, tender, odorant, and juicy, it must rest a few minutes so that the juices in the center migrate toward the dry periphery.

Then it must be cut with a good knife that will not maul it, allowing the juices to be retained within.

Other countries, other customs: the English are renowned for their boiled meats but also for their well-done roasts. Across the English Channel, a roast remains respectable when the temperature at its center reaches 60° (140°F) or even 80°C (176°F). In France, the temperature at the center of a roast sometimes does not even attain 30°C (86°F).

How Does a Roast Cook?

Let us recall the paragraphs devoted to cooking and review one by one the various kinds of heat transfers in order to see which one applies to roasts. We will consider three of them: radiation, convection (the hot air ensures the distribution of heat in the oven), and conduction (within the meat).

How does a roast cook? First of all, note that the thermic conductivity of meat diminishes considerably with temperature. Meat conducts heat very poorly (it is an insulating material) at low temperatures (at 0°C [32°F], for example), but it becomes a better conductor at higher temperatures. We can examine the consequences of these properties by analyzing how a Christmas turkey cooks.

THE CHRISTMAS TURKEY

How to cook a turkey correctly? A turkey's sphericity holds some interest for physicists, who know how to calculate heat transfers within bodies with simple forms. In 1947 Horatio Scott Carslaw and John Conrad Jaeger studied the relationship between the radius of an ideal, spherical turkey and cooking time. They assumed that a turkey was a mixture of water, fats, and proteins in the ratio of 60/20/20, and they sought the optimal cooking temperature.

Why does the cooking temperature matter? When a turkey is roasted, its fibers contract until, at about 70°C (158°F), the individual muscle cells begin to deteriorate. During heating, the weakest bonds between the atoms of certain molecules are broken, so that the proteins are denatured. These long threads, folded back on themselves in specific configurations, unwind and move in all directions. Because the proteins can thus come into contact with one another, they bond and coagulate. The meat hardens, but not very much.

When the cooking is too prolonged, the water that remains bound to the proteins is released, and the meat becomes tough. Conversely, we have often seen that, the longer the cooking, the more the rigid network of collagen is broken down. In short, the cook who roasts a turkey must find a compromise in order to break down the collagen and at the same time avoid letting the proteins dry out and toughen after they have coagulated.

COOKING CALCULATIONS

Since it is juicy, tender meat that we want, it is clear why there is no question of opening the oven while the meat is roasting. The water vapor that is released in a limited quantity could escape and then be replaced by the vaporization of a certain quantity of the juices. Opening the oven dries out the turkey. Neither, however, should one humidify the oven before putting the turkey in. In the presence of too much water, the surface water cannot evaporate, and the skin will not get crispy.

Having thus resolved the problem of the surface, the serious problem of tenderness within remains. We cannot disappoint our guests, who fear the proverbial dryness of the Christmas turkey.

Since tenderness results necessarily from the deterioration of the connective tissue, let us consider this tissue. It principally contains three kinds of proteins: collagen, already discussed many times, reticulin, and elastin. Neither reticulin or elastin are notably altered by the heat of the oven, but the triple helixes of the collagen molecules can be broken up and form gelatin, which is soft when it is in water, as we all know.

Calculating the cooking time requires some skill, because the denaturation of the collagen and the coagulation of the muscle proteins (actin and myosin, mainly) take place at different temperatures and different speeds in the different parts of the turkey. It is necessary to know that the temperature of 70° (158°F) is essential for transforming the collagen into gelatin and tenderizing the muscles. But the longer the turkey remains at a high temperature, the more water it loses and the more its proteins risk coagulating. The optimal cooking time, consequently, is the minimum time it takes to attain the temperature of 70°C (158°F) at the center of the turkey.

Thus defined, the problem is simplified because physicists are experts on heat transmission in different materials, even animal tissues. They make their calculations assuming the temperature is homogeneous and the animal is cylindrical, homogeneous, and so on, and they end up with precise if complex results involving the radius of the animal, the coefficient of the thermic diffusion, the temperature of the turkey, and the temperature of the oven.

Fortunately, there is a simpler means for calculating the cooking time for turkeys. We can apply the Fick law, which stipulates that, for the center of the turkey to attain a given temperature, the time of heating (t) is proportional to the square of the radius of the turkey. Since the mass (M) of a sphere is proportional to the cube of its radius, we can determine the cooking time by applying the simple formula $t = (M/M_o)^{2/3}t_o$.

My friend Peter Barham, a physicist and molecular gastronomer from Bristol, has calibrated this equation and calculated the following values:

- at 180°C (356°F), a turkey weighing five kilograms (11 pounds) must cook two hours and twenty-five minutes, and a turkey weighing seven kilograms (15.43 pounds) must cook three hours;
- at 160°C (320°F), a turkey weighing five kilograms (11 pounds) must cook three hours and thirty-five minutes, and a turkey weighing seven kilograms (15.43 pounds) must cook four hours and thirty minutes.[1]

One Can Become a Cook, but One Is Born a Roaster

Brillat-Savarin's aphorism paraphrases an old Latin adage: *poeta nascitur, orator fit* (one can become an orator, but one is born a poet). I will end this chapter by attempting to prove Brillat-Savarin wrong; to that end, let me succinctly

---

1   There are other possibilities, of course. You can put the turkey in the oven at a low temperature in order to give the collagenic tissue time to dissolve. Then, when the meat is tender, you can grill the surface, thus completing the preparation of the dish.

analyze the classic rules of roasting, adherence to which will allow anyone to roast as to the manner born.

You want to obtain meat that is just right, that is to say, nicely browned, every bit of the flesh enhanced by the cooking, and oozing with juices when carved. To retain the juices, shorten the cooking time by placing the roast in a preheated oven, as hot as possible (too hot, and the meat will char).

As in the case of stew, season the roast only after completely cooking it, a few moments before taking it off the spit or out of the oven, because, once on the meat, the salt will draw the water from it, hamper browning, and dry out the inside (through the phenomenon of osmosis). Likewise, do not add pepper until the end of cooking. Pepper, when overheated, gives meat a strong, unpleasant taste.

Lard the meat with strips of pork fat or wrap it in a caul or strips of bacon; this enhances the browning by releasing fat, which will aid the transfer of heat and protect against overheating.

Finally, remember that the only true roasting takes place on a spit, because this is the only way to ensure that all parts have equal contact with the heat. Meat in a metal pan, for example, cooks more quickly at the points of contact than elsewhere.

THE STUFFING AND THE SAUCE

While the turkey slowly roasts, let us examine how the stuffing we have put inside it cooks.

Remember that the stuffing, placed where the temperature is lower, will have a harder time cooking than the turkey. It is especially for this reason that stuffing is often forcemeat mixed with egg. Prepared in this way, the meat filling the turkey forms one coherent mass. The 70°C (158°F) attained at the center at the end of the cooking time is high enough to coagulate the egg and bind the various components in the stuffing.

Additionally, in order to avoid burning the outer layer of the meat, let us baste it for about three full minutes with the melted fat (not liquid, which would soften the outside crust). This forms a screen, stopping some kinds of radiation. Sometimes the roast can be protected by covering it with paper coated with oil, but basting is more effective. Naturally, we should not begin to baste before a crisp crust has formed on the surface.

The fat that drips over the meat falls back into the bottom of the roasting pan (or, if one is using spit, into a pan set below the roast to collect the drippings), bringing with it a bit of the browned juices of the meat.[2]

At the end of cooking, these delicious juices are recovered in what is called a glaze. Once the pan is removed from the oven, the meat is lifted out of it, a little boiling water is poured into the pan to dissolve these juices, the mineral salts from the meat, and the gelatin that emulsifies the fat. With a little whisking, a thickened sauce with a good consistency is quickly obtained.

If the sauce does not emulsify and remains in two phases, it can easily be homogenized with the help of gelatin.

Finally, remember that, depending on the kind of roast, the deglazing of the dripping pan can also be done with wine, milk, diluted cream, cognac, whatever you please.

---

2 Once again, this is not really a matter of caramelizing. Caramel results from cooking sugar, whereas the brown juices of meats are formed by Maillard and other reactions.

Deep-Frying

Why Is It Necessary to Fry in So Much Oil?

All cooks know that frying is cooking through contact with hot oil. They know that this operation creates a golden crust, but they distrust heated oil, which spatters, makes the kitchen greasy, and produces the taste and smell of burnt fat as it gradually darkens. And that is why some physicians condemn deep-frying on dietary grounds.

The makers of household appliances have already overcome the first two inconveniences of deep-frying by inventing deep-fryers equipped with a filter and in which the operation takes place in a closed vat. Is there a way to overcome the last inconvenience, of reconciling the pleasures of deep-frying with the concerns about health or even one's figure? How to fry well? What oil to use for deep-frying?

The principle of deep-frying is simple. The heated frying pan transmits its heat to the oil, which can rise in temperature well above the 100°C (212°F) maximum attainable with water. Brought to these very high temperatures, the cells on the surface of the foods to be fried, in drying, produce the characteristic crispness of good deep-fried food. Have you noticed that the surface of fried food seems dry? That is because the surface moisture, brought suddenly to a temperature higher than 100°C (212°F), has evaporated.

How to obtain good fried foods? By using oil as hot as possible, because if the crust is not formed very quickly, the oil penetrates the food. A dramatic release of vapor bubbles from the food's surface indicates that the frying process is going well.

Moreover, the initial temperature must be increased according to the size of the food. As it heats, the food cools the oil in which it is placed. A large piece of food cools the oil more than a small piece. Since the maximal temperature of the oil is limited, a good solution for cooking a large quantity of food is to use a large quantity of oil, in which a large quantity of heat will be stored.

It would be a serious mistake to use a small amount of oil for frying on the pretext of worrying that it will saturate the food. On the contrary, the food will fry in oil that is too cool to sear it and thus will become a horrible oil sponge.

It is equally important to realize that oil cannot withstand excessive heat. Just as overheated butter blackens and burns, too-hot oil deteriorates. A good deep-fry cook must remember that all oils do not have the same capacity for withstanding heat.

Try this experiment. Take oil and heat it quickly. It will eventually release a strong, sharp odor and a pungent smoke. A compound called acrolein is the source of this acridity that the cook must avoid. We can now turn to the farm-produce industry, which is working to produce special oils for deep-frying, that is to say, oils that have as high as possible a smoking point.

Why Must the Deep-Fry Oil Stay Clean?

Even the best oil can only produce good fried food if it is treated with care.

We know that reused oils smoke as soon as the temperature rises even the slightest bit. They have lost their frying properties because they have gradually become full of little particles, of meat, for example, which cook at 70°C (158°F) and blacken above that temperature, releasing acrid compounds. Moreover, the oil itself generates compounds that further increase its degradation. The solution is imperative: if an oil must be reused, it must be filtered in order to remain clear.

This same phenomenon of protein carbonization prevents the use of butter, without some preparation, for deep-frying. At the temperature of 45°C (113°F), butter melts; at 100°C (212°F), it sputters (because the water it releases evaporates); then, at 120°C (248°F), it decomposes unless someone has taken the precaution of clarifying it.

Even though it is simple, clarifying butter is an operation lost among the household arts. What does it consist of? Eliminating the proteins (especially casein) that the butter contains, in order to obtain a fatty substance as pure as possible that can withstand a good heating without turning black. Decomposing at a low temperature, the proteins in butter darken and impart a burnt flavor at the same time that they prompt the decomposition of the butter's lipids.

And, say what you like, deep-frying done with the aid of a good clarified butter is a true gastronomic pleasure. (This same clarified butter will be very useful in many other preparations, such as grilling.)

How to set about clarifying butter? It is a matter of placing the butter in a saucepan and heating it a long time and very gently. After about thirty minutes, the caseins precipitate. The supernatant (the clarified butter), which can be recovered by simply pouring it into a container where it can be stored, retains the sapid and aromatic qualities of the original butter.

Why Must the Food for Frying Be Dry?

We have seen that there must be an abundant quantity of frying oil, because the thermal inertia of a heated body is proportional to its mass. Cold pieces of food introduced into heated oil cool the oil less if there is a large amount of it.

Let us also not forget that the pieces of food should be cut small, if possible, so the inside has time to cook before the surface molecules begin to burn.

And, finally, let us remember that foods placed in hot oil must be dry. First of all, it would be a useless loss of heat if the oil first had to evaporate the water on the surface of the food before carrying out the actual frying. Second, one can avoid spattering fat if the foods for frying are dry. When water is suddenly immersed into oil at a temperature much higher than its vaporization temperature, it is transformed very quickly into vapor and, by expanding so violently, spatters fat everywhere.

What to Fry?

I will not launch into a tiresome list, but I do want to note that the flaky, crispy consistency, the golden color, and the characteristic flavor of fried foods are due in part to the coagulation of proteins and the caramelization of glucides (sugar and starch) over the course of the frying. That is why potatoes were predestined to be deep-fried: they consist, on the surface, of sugars and starch that are favorably transformed.

Nevertheless, the preparation of fried potatoes can go wrong when too many or too large pieces are placed in the oil. The fat (which smokes at about 190°C [374°F]) is cooled to 130°C (266°F) and remains at that temperature, so the potatoes do not cook.

For perfect french fries, place the potatoes in a large quantity of hot oil. After about five minutes, turn up the heat to increase the oil temperature, so that a crisp crust forms. Then, as soon as you remove them from the oil, blot the french fries with paper toweling. This method takes into account the pressure (yes, the pressure!) inside french fries. When measured, the pressure can be observed to increase gradually as the water in the potato evaporates as they cook. Thus a french fry has a crisp outside, some purée, and a lot of vapor inside (cutting one open will reveal this). When the french fries are taken out of the oil, the pressure decreases because of the recondensation of the vapor into water, which sucks in the oil on the potato's surface. Immediately blotting the french fries reduces the amount of oil they absorb.

For foods that do not contain glucides on the surface, bread crumbs are a good solution, since they come from flour, which is essentially composed of glucides. However, since bread crumbs will not stick to meat, for example, you must coat such foods in beaten egg before applying the bread crumbs. The egg binds the bread crumbs to the meat, and it also provides proteins that react chemically with the sugars through Maillard reactions (again!).

You can improve the process and prevent the crust that forms from coming detached by first dusting the meat with flour, then dipping it in egg, and finally applying the bread crumbs. The coagulated layer of bread crumbs will stick to the meat because of the starch that is formed. This method is even more effective if you first pierce the food with a fork. The egg and flour penetrate these holes and further anchor the fried coating to the food.

Sautés and Grills

Braising with Very High Heat

Strictly speaking, to sauté is to cook meat, fish, or vegetables in a fatty substance over high heat, uncovered, and without adding any liquid. In practice, however, for large pieces of food especially, this first phase of cooking must be followed by more gentle cooking, in an open pan, so that the odorant molecules in the vapor can add the finishing touches to the initial sauté. A true sauté differs from braising in that, with an uncovered pan, no vapor limits the cooking temperature. As with deep-frying, cooking takes place at a temperature higher than the 100°C (212°F) of boiling water.

For sautés, the fatty substance is of primary importance. To obtain good sautéed meat, vegetables, or fish, clarified butter is essential because, in addition to its flavor, it can withstand a temperature higher than can natural butter without burning. Can higher temperatures also be attained by mixing oil and unclarified butter? Our experiments did not confirm this old wives' tale.

For grilling as well, the cooking takes place at high temperatures, although without oil. The meat rests in direct contact with the grill. To improve the contact and transfer of heat, the meat can be brushed with a little oil or clarified butter.

Many culinary works claim that the superficial caramelization of the meat's proteins, in forming a crust, forms an impenetrable layer that traps the nutritive juices. (We have seen that there is no caramelization involved and there is no impenetrable layer; I will come back to this in the next section.) Additionally, books advise not to salt or prick the meat so as to avoid the loss of juices. Do you recall that these same instruction were given with regard to roasts?

Good Sense Gone Wrong

Adding salt is certainly to be avoided in some cases, because the phenomenon of osmosis causes the juices to escape the meat when muscular fibers are cut and open, and pricking the meat is harmful because it creates channels from which the juices can leak out. But the impermeable crust is a myth for which the German chemist Justus von Liebig (1803–1873) is responsible.[1] In the nineteenth century, Liebig understood that heat coagulates the proteins on the meat's surface. He extrapolated, however, when he assumed that the coagulated crust trapped the juices. The idea that cooking with a hot flame can cauterize meat and limit the loss of juices, though never proven, traveled rapidly to England, then to the United States, and finally back to France, where it has reigned in error until very recently.

Many observations carried out by Harold McGee in Palo Alto, California, demonstrate the falseness of Liebig's hypothesis.[2] First of all, a grilled steak sizzles while it cooks. That is a sign that liquid—the juices—is escaping the meat and vaporizing. The sizzle is the sound of the vapor spurting out.

Second, even if the steak is removed from the pan, the plate that receives it is soon filled with juices. These juices escape the meat as soon as it is finished cooking. Thus the supposedly impermeable layer is hardly that.

Third, if the pan is deglazed with wine, for example, it is to dissolve the juices that have escaped the steak during the cooking process and caramelized deliciously.

---

1   Before his career as a remarkable chemist and pioneer in the chemical analysis of organic products, Liebig learned to speak French in the kitchen of the duke of Hesse-Damstadt. He drew on his "culinary education" when he studied chemistry in France, as a comparison of his subsequent publications and old French culinary texts makes clear. After moving from Giessen, where he created a prototype for modern chemistry laboratories, to Munich, Liebig began to apply the information he gathered from his chemistry experiments to cooking. In particular, he is given credit for the scientific discovery discussed in the text.

2   Not content with having published his astonishing book *On Food and Cooking*, Harold McGee has presented these and other culinary/scientific investigations in a second very interesting book, *The Curious Cook* (North Point, 1990).

Fourth, that vapor that is released throughout the cooking: what is it, if not the juices vaporizing?

All in all, it seems obvious that juices leave meat as it cooks, even if the surface is seared at the very beginning of the cooking process. By contracting the connective tissue that surrounds the muscle fibers, the cooking process prompts the expulsion of the meat's juices.

So how to retain the most juices possible in grilled or sautéed meat? One solution consists of not overcooking it, naturally. The less the connective tissue is contracted, the less juice is expelled. A second solution consists of cooking with high heat. In this way, the meat cooks rapidly, and the juices do not have time to escape the meat in very significant amounts. Third, salting and pricking the meat should be avoided, for the reasons previously shown. Finally, the grilled meat should be eaten without delay, as soon as it is cooked. In this way, the juices will not have time to leak out onto the plate.

As an aside, if a sauce is served with the grilled meat, it must be thickened a bit more than may seem necessary, because it may be diluted by the juices that will inevitably leave the meat and run out onto the plate.

Even More Tender

Between Tough and Putrid

Very fresh meat is tender, but fresh meat is tough; gradually it becomes tender again, and then it rots. How to conserve it in that precarious state in which it does not require excessive chewing but neither does it release an unbearable odor, revealing an unhealthy degree of putrefaction? Our ancestors invented many processes for long-term conservation: smoking, salting, drying. But today's cooks can get meat anytime at all from the neighborhood butcher, who sells cuts that are aged for exactly the right amount of time. They no longer have to solve the problems of long-term conservation. Beginning with products aged under supervision, their chief goal is to obtain meat that will be tender after cooking.

Our consideration of stews showed us the importance of lengthy cooking in a liquid, in order to break down the collagen fibers that toughen meat. Other processes produce the same result. Hanging, marinating, and "proteasizing" are appealing for reasons it would be a shame to overlook.

By the Neck or by the Beak?

Let us read Brillat-Savarin:

> Above all other feathered game should come the pheasant, but once again few mortal men know how to present it at its best.

A pheasant eaten within a week after its death is more worthless than a partridge or pullet, because its real merit consists in its heightening flavor.[1]

How can we attain the summits reached by the master? Ask all around. You will hear that hanging game is an abominable operation, that our forefathers ate putrid meat, that, after being suspended by its neck, the pheasant was consumed when it fell, its head detaching from its body because of rotting. Is that the last word on the subject? Would we be here today if our ancestors willingly poisoned themselves on rotten pheasant? And who among us has seen, with his own eyes, the hanging of a pheasant by its head?

Bibliographic research has shown me that hanging pheasant does not follow any absolute rule but that good sense is essential. First of all, yesterday's great cooks did not recommend hanging pheasant by the neck, or even by the beak, but by the tail feathers. Being generally a heavy bird, a pheasant falls well before it rots.

A second precept is that the animal must be hung with its feathers still on, which protects it from insects and other small pests who would threaten our feast. Finally, the length of the hanging depends on the temperature and the weather. Just as we see fish turn when they are not cleaned and there is a storm, a pheasant can only hang for two or three days when the air is humid but can remain in a cool draft for six days when the weather allows. Supposedly, during this operation, a special juice that is present in the shaft of the feathers is reabsorbed into the flesh. This claim deserves an experimental study.

Brillat-Savarin, whose colleagues, it is said, took offense at the odor of hung meat that accompanied him (supposedly he put pheasants in his pockets to age them), wrote that "the pheasant is an enigma whose secret meaning is known only to the initiate."[2] How to become an initiate? How to cook a pheasant with the attentive care necessary to take it to the point where it surpasses a good chicken?

---

1   Jean Anthelme Brillat-Savarin, "Meditation 6," sec. 39 in *The Physiology of Taste; or, Meditations on Transcendental Gastronomy*, trans. M. F. K. Fisher (New York: George Macy Companies, 1949; reprint, New York: Counterpoint, 1999), part 1, p. 87.

2   Idem, sec. 12 in ibid., part 2, p. 374.

According to Grimod de la Reynière, "the pheasant is done on a skewer, wrapped in paper, well buttered. The paper is then removed to give it good color; then it is served in a verjuice sauce, with pepper and salt." Today, instead of the verjuice, made from unripe grapes, a slice of lemon, salt, and pepper can be substituted.

How Many Days for Marinades?

If hanging is good for pheasant and its feathered cousins, marinade is more suitable for large, furry beasts, like wild boar (which is often tough), mutton, and beef.

The process is simple. The meat rests in a mixture of wine, oil, vinegar, spices, various condiments, and a few vegetables (this mixture can be cooked beforehand). With time, the meat becomes tender and flavorful. Subsequent cooking completes the dish, be it grilling, roasting, cooking it in the marinade itself, in short, whatever you prefer.

What are the principal elements for a marinade? Vinegar, flavors, time.

Vinegar is an acid that attacks the connective tissue and breaks it down. That is one reason it was thought that the meat gets tender, but not the main reason. From our laboratory experiments, we concluded that meat becomes tender in a marinade because, while it is protected from putrefaction, the muscular fibers age and protein aggregates are slowly dissociated, just as when butchers age meat in their special refrigerators. Previously, we had thought that tenderization occurred where the flesh had been in contact with the marinade. But our more recent experiments have shown that a marinating solution penetrates meat to a much more limited extent than our model systems had suggested, in which this diffusion occurred at the rate of about ten millimeters per day.

Other experiments, more culinary in nature, lead to impressive results. A marinated roast pork can be taken for a leg of very young wild boar; marinated mutton can pass for venison.

Whether you wish to fool your guests or not, serve marinated meats with red currant jelly: it's delicious!

Pineapple Power

Having explored various methods of tenderizing meat, my esteemed friend Nicholas Kurti, whom I have already mentioned, during a March 14, 1969, meeting of the Royal Institution to which the BBC had been invited, demonstrated that an injection of fresh pineapple juice into a pork roast resulted in absolute tenderization.

Another English fad? Not entirely, because first, although Nicholas Kurti was a professor of physics at Oxford and a member of the very old and venerable Royal Society of London, he was Hungarian by origin. A longtime record holder for the lowest temperature ever reached (a millionth of a degree below absolute zero, that is, about 273 degrees below the temperature at which water freezes), Nicholas Kurti was a passionate cook. With his public experiment, he wanted to demonstrate the power of the enzymes in pineapple juice and confirm experimentally a method extolled by the Aztecs.

Enzymes are molecules that promote various reactions in live bodies. They are found in all living cells and notably in fresh pineapple, papaya, and fig juice, among other plant products. The specific enzymes found in pineapple, papaya, and fig (bromelain, papain, and ficin, respectively) have one peculiarity: they are proteolytic, that is, they break down proteins. Now, meat, as we have seen on several occasions, is composed of many proteins; collagen, especially, responsible for meat's toughness, is a protein.

Nicholas Kurti demonstrated how to put the useful properties of these enzymes to work in preparing meat. He squeezed a fresh pineapple, placed the juice in a hypodermic syringe, and injected the pineapple juice into a pork roast (in just one-half in order to compare the results of the enzyme action). He let the roast rest for a few minutes, so that the enzymes would have time to react. Then he put the roast in the oven and let it cook for less time than necessary to cook the untreated half thoroughly.

Taking the roast out of the oven, he cut it into slices. The half that had not received the pineapple juice was still the pink color characteristic of undercooked pork, even though the meat was covered with a crisp crust. By contrast, the meat on the other side was almost reduced to purée. Naturally, the meat had a distinct pineapple taste, but isn't there a recipe for pork with pineapple?

Medicine and Cooking

"From your foods, you will make your medicine," said Hippocrates. While we wait for modern medical nutritionists to define for us the perfect foods to ensure our health and longevity, let us borrow one of their instruments: the syringe.

This tool, used by Nicolas Kurti for his pineapple juice injection, can also improve the marinating process.[3] While the meat marinates, draw off the marinade at regular intervals and use the syringe to inject it into the meat. The results are superb because the marinade works from the inside out, and the preparation time can thus be shortened.

In addition, a number of cookbooks mention that marinated meat must not be roasted, for fear of drying it out. That is correct, according to my experience, but by injecting the marinade into the center of the meat, you can avoid that danger.

3 Nicolas Kurti was not the first to use a syringe for this purpose. "Intrasauces" were introduced as early as the 1920s by Dr. A. Gauducheau, who published wonderful articles describing them in various serious scientific journals in 1931.

Salting

Why Must Infants Not Be Fed Sausages?

Those nitrates that ecologists condemn for polluting streams and rivers are present in foods preserved with salt. Potassium nitrate, that is, saltpeter, has been used in this way empirically since the Middle Ages, even since Roman times. In 1891 the biologist H. Polenski demonstrated that bacteria transform saltpeter into nitrite in meat. Then in 1899 came the discovery that the characteristic color of salted products was due to these nitrites and not to the nitrates themselves. In 1901 the biologist John Scott Haldane found that this color resulted from the combination of the chemical group NO with the meat pigments. Finally, in 1929, nitrites were observed to inhibit the development of bacteria. Today, the description of the process is complete: salting, with the use of saltpeter, is an effective conservation method because the nitrate ions of the saltpeter are transformed into nitrite ions, which kill bacteria.

Unfortunately, nitrites are certainly not lacking in toxicity for humans as well. They react with the amino acids that make up proteins and form carcinogenic nitrosamines. Babies, especially, should not absorb nitrites, because these compounds are oxidants. They transform the hemoglobin in the blood into methemoglobin, which no longer transports oxygen. Adults possess an enzyme called methemoglobin reductase that retransforms methemoglobin into hemoglobin, but infants, who do not yet have this protective enzyme, must wait to indulge in sausage, dried meats, and the like.

How Do We Dry Meat Using Salt?

Though nitrited salt is available commercially, we should nevertheless remember that nitrites are not crucial for home salting. A well-implemented brining and drying process will suffice. The salt in a brine acts according to the phenomenon of osmosis already discussed. When a piece of meat is placed in a terrine with a little water and a lot of cooking salt, the water in the animal cells tends to leave the meat until the concentration of salt inside and outside the cells is equal. The salt does not enter the cells, but the water, small molecule that it is, is very mobile.

Thus drained, the meat hardens on the surface, and in this waterless meat bacteria have trouble developing. Why must a little water be added to the terrine? Isn't the cooking salt alone sufficient? With a bit of water, the meat is entirely soaked, so that contact with the salt is improved.[1]

After undergoing this treatment for some time, the meat is removed from the brine and dried. For a successful drying operation, it is advisable to place the meat in a dry, well-ventilated spot. An uninsulated attic with good ventilation serves well, as does a cool, dry cellar. The meat dries, and, after some time, it can be consumed . . . with that great pleasure derived from slowly cured, long-awaited foods.

---

1   Cooks may want to know that this process also works very well using sugar. Sugar dissolves better in water than salt does, so it draws water out of meat more efficiently, and since it does not enter the meat itself, there is no danger of producing sweet meat.

Microwaves

Cooking with Internal Vapor

Cooked in a microwave, beef is rejected by taste testers, who find fault with its grayish external color, the uniformity of its internal color, its toughness, its lack of succulence, and its bland taste. And they are right. Microwaves penetrate into the mass of the foodstuff for several dozen millimeters before being absorbed by the water molecules. These molecules are heated, then vaporized. The temperature never goes above 100°C (212°F). Now, as we have seen, heating in this way is fatal to meat, which must be heated intensely to achieve the browning produced by Maillard and other similar reactions.

On the other hand, microwaves are good for cooking eggs, for example, in which the proteins coagulation begins at 61°C (141°F). Placed in a bowl without an ounce of fat, an egg will cook rapidly; its taste is acceptable, and the figure benefits. Scrambled eggs, soft-boiled eggs, omelets, and even soufflés can be cooked in this way. Microwaves are useful for fish as well, because they efficiently heat the poaching liquid over which the fish is placed. Similarly, vegetables can be cooked in boiling water, heated by microwaves.

Where Do Microwaves Act?

Let us review the basic principles in order to really understand where microwaves act. Inside a microwave oven is a device called a magnetron that emits electromagnetic waves (that is, vibrations in space analogous to light or to radio waves but with a different wavelength) with a frequency equal to 2400 megahertz. At each point in space crossed by a microwave beam, the electromagnetic field oscillates 2400 times a second.

Without safeguards, such waves would heat the water in our bodies, and we would boil. Thus the waves are directed by an aluminum tube just inside the oven, and they are sealed within the oven (metallic grating in particular, like the kind used to reinforce microwave oven doors, stops microwaves).

When food is irradiated with microwaves, the waves interact through their electrical field with electrically asymmetric molecules, such as water molecules. The energy given to these molecules is transformed into motion, and the movement of these agitated molecules disturbs the other, unagitated molecules, so that the mass is put into motion, that is to say, heated. Gradually, the agitated molecules are calmed down by colliding with the surrounding molecules, through their random movement. Since most foods contain large quantities of water, they are heated because this water becomes agitated, and it is especially the parts of the food containing the most water that are heated the most. Hence the recipe for *canard à l'orange* given at the start of this book.

A Few Questions and Answers

Why have the manufacturers of microwave ovens adjusted the frequency of microwaves so that they are a bit lower than the frequency at which water best absorbs these waves?

Because if we want the inside of foods to be cooked as well as the outside, the microwaves must not be absorbed immediately by the outer layers of the food. If the water on the surface absorbs only some of the microwaves, the rest will permeate the food, where another share will be absorbed.

Why do salted foods heat more quickly than unsalted foods in a microwave oven? Because salt contains ions, and the water molecules that hydrate these ions, by surrounding them, heat more quickly than isolated water molecules.

Why should we not try to heat oil in a microwave oven? Because triglycerides have no chemical groups that can interact efficiently with microwaves. This can easily be demonstrated by putting two glasses, one filled with water, one filled with oil, in a microwave oven. When the water comes to a boil, the oil will still be cold.

Why does meat cooked by microwaves become grayish-brown? Because the temperature stays below 100°C (212°F); thus the oxymyoglobin is not denatured and retains its color.

And, to end with something sweet, remember that caramel can be prepared quite easily in a microwave oven. Take a small bowl, place sugar and a bit of water in it, and heat. Caramel rapidly results without any trouble whatsover.

Vegetables

COLOR AND FRESHNESS

A Matter of Water

Vegetables, the jewels of the kitchen! Did they not give their names to the great Roman families? Fabius, in honor of *faba*, or *fève*, the broad bean; Lentulus, in honor of the lentil; Piso, in honor of the pea; Cicero, in honor of the chickpea.

Vegetables must be eaten fresh to be good. The soil in which they were cultivated, the climate that brought them to life will sing in one's mouth . . . if they are not mangled in the cooking process. Cooking them is a delicate operation. How long must they cook to become sufficiently tender? Must they be tossed into cold or hot water? Must the cooking water be salted? How to retain their bright colors, which seem to be the mark of their freshness?

Before I launch into an examination of this last question, let me recall that a very fresh vegetable is generally tender, and cooking is not of great value to it. On the other hand, for certain older or even dried vegetables, like lentils, rehydration is essential.

In these two cases, the cooking methods are very different, since the object in the first is to retain the emollient moisture of the vegetable and in the second to reintroduce moisture that has been lost.

How Do We Avoid the Discoloration of
Green Vegetables When Cooking Them?

The intense green that vegetables acquire after cooking for a few seconds in boiling water results from the release of gases trapped in the spaces between the vegetable cells.

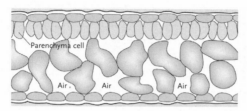

Generally, these pockets of air act as magnifying glasses that highlight the color of the chloroplasts, the green organelles that are responsible for the transformation of carbon dioxide into oxygen.

Vegetables, however, are usually cooked longer than a few seconds, thus destroying the atmosphere that shows these vegetables in their best light. Spinach cooked too long turns brown, sorrel as well; leeks lose their greenness, and so on. How to retain that appetizing color?

The cooks of antiquity were the first to make advances toward explaining this phenomenon. They observed that green vegetables remained very green when saltpeter or ashes were added to cooking water. Why?

When a green vegetable is heated, some of its cells burst, releasing various organic acids. The hydrogen ions of these acids react with chlorophyll molecules (which contribute to the green color of green vegetables) because these molecules contain a large square chemical pattern, the porphyrin group, at the center of which is a magnesium atom. Now, the hydrogen ions have a bad habit of taking the place of the magnesium ion in this porphyrin group and of thus transforming the various kinds of chlorophylls into pheophytins, which absorb different components of light. Instead of retaining all the light rays except those of the color green, pheophytins reflect a mixture of wavelengths that produce the perception of a horrible brown.

But from this analysis emerges a solution: not heating the vegetables for too long, so that the magnesium will remain in its chlorophyll cage.

A few corollaries are equally essential. To retain the color of green vegetables, avoid lidded earthenware pots and opt for steaming, because if they are not immersed in water, the vegetables are not in contact with the hydrogen ions. If vegetables are cooked in water, large quantities of water should be used. Finally, adding vinegar to the cooking water for green vegetables should be absolutely avoided, as it will enhance the bad effects you wish to avoid. Be aware, too, that many juices from fruits are very acidic (and that the acidity one perceives can be hidden by sugars).

Naturally, inventive cooks have thought of cooking green vegetables in the presence of salts, which provide ions that can occupy the positions hydrogen ions would otherwise take. That is why green vegetables were cooked in copper pans, called "regreening pans," and why, later in history, copper salts were used; with these methods, the green remained intense . . . but the vegetables became toxic. Indeed, a law prohibited the practice of adding copper salts in 1902. More recently, processes using zinc ions have been patented.

Adding a base to the cooking water in order to neutralize the acids as they form has also been considered. This solution was already familiar to the Romans. Apicius, famous for his gastronomical extravagances, wrote, "*Omne holus smarugdinum fit, si cum nitro coquantur*" (All vegetables will be the color of emerald if they are cooked with niter).[1] The same effect occurs with ashes, where potash is present. Alas, niter, or saltpeter, and potash ruin the taste.[2]

How Long Must Vegetables Be Cooked?

Do not hope for a global response to such a question. Fresh asparagus will cook for less time than asparagus kept for a day or two after picking. And regardless

---

1   There is a story that Apicius once chartered a ship to Tunisia to look for shrimp—big ones, he had been told, as big as one's hand. When he arrived and saw some specimens, he departed, disappointed, without even touching land.

2   Another practice, recommended in old French cookbooks, involves cooking vegetables in a potash solution obtained by soaking ashes in water. And it works! But again the flavor is horrible. More recently, bicarbonate of soda has been used. That seems the best solution, although it, too, produces a bland, basic flavor rather than a bright, acid one.

of freshness, asparagus will not take as long to cook as potatoes. Still, as is so often the case, an analysis of the problem can guide us in our culinary transformation operations.

The objective is to tenderize the vegetables, the cells of which, unlike animal cells, are each protected by a hard, fibrous wall. Weakened by cooking (the cellulose is not altered chemically, but the pectins and the hemicellulose are), these walls becomes porous, and as their proteins are denatured, they lose their ability to regulate the movement of water from the interior of the cell to the exterior, and vice versa. Water can pass through the walls, while larger molecules are blocked.

We know that when we put vegetables into unsalted water, they swell because the water enters the vegetable cells as a result of osmosis. On the other hand, if the cooking water has too much salt, the vegetables harden (especially carrots), because the water does not enter the cells to reduce the salt concentration in them—the contrary!

The Mystery of Dried Vegetables

The case of dried vegetables (lentils, etc.) is a little different, because the objective there is to reintroduce the water lost in drying them. As I just noted, the cooking water must not be salted. Nevertheless, this precept is not enough, and cooks have perfected a precise methodology for obtaining good results.

The first operation should be a soaking, the aim of which is to soften the external layer of the vegetable and facilitate the subsequent cooking. Often, two hours of soaking is enough to obtain a wrinkled skin. Warm water seems preferable to cold water, because the soaking is thus accelerated. The soaking water is then replaced for cooking.

The cooking water must not be calcareous, cooks say, because if a layer of calcium settles on the skins of the vegetables, it will harden them and prevent them from cooking. Authors like Madame Saint-Ange recommend adding bicarbonate of soda when the water is calcareous. In fact, no layer of calcium forms, but calcium should be avoided nevertheless because it acts as a cement between pectin molecules in the vegetable cell walls, hardening them rather than promoting softening. Madame Saint-Ange was right to recommend bicarbonate of soda. It has two benefits. First, the calcium is precipitated so that it cannot bind the pectins. Second, the water becomes basic, contributing to the pectin separation (we shall witness this effect again later, with regard to jams).

It is also specified that the cooking be gradual. This makes sense in principle, because cooking too rapidly from the outset cooks the exterior part too much, turning it to mush before the center of the vegetable is soft. Likewise, adding cold water if the cooking water boils away should be avoided. The sudden thermal discontinuity can explode the vegetable skins, thus releasing their contents into the cooking water.

Do Carrots Risk Losing Their Color in Cooking?

If the cook is careful, no carrot will ever lose its color. We must understand that the color of vegetables comes from various pigments: chlorophylls (green to blue pigments), carotenoids (yellows, oranges, and reds), and anthocyanins (reds, purples, and blues). If green vegetables are green, it is because they contain chlorophylls. If carrots are orange, it is because they contain, especially, carotenoids.

Now carotenoids, soluble in fat but insoluble in water, are little altered by boiling water. Normally, carrots remain brightly colored (the same is true for tomatoes, though their color is mainly due to lycopene, not carotenoids). In other words, carrots are easy to cook . . . so long as a pressure cooker is not used. The pressure that builds in a pressure cooker alters the carotenoid molecules, which then lose their color.

How Do We Cook Potatoes?

Potatoes are made of cells that contain granules of starch. These starch granules become soft, inflated, and jelled when they are immersed in water at temperatures from 58° to 66°C (136° to 150°F). The perfectly cooked potato is full of these inflated, tender granules, all of which have uniformly reached the temperature of 66°C (150°F).

Thus sautéed potatoes are better when they have been cooked in water for a few minutes and acquired a jelled outer layer. During cooking, this layer prevents the starch granules from absorbing too much oil, while the external surface can be heated to 160°C (320°F). The starch contained there deteriorates and reacts, as we saw in the chapter on deep-frying, giving way to a crispy, golden casing.

Can a Dish Containing Vegetables Be Reheated in Butter?

Reheating vegetables in butter is often a mistake, because the butter will make the sauce oily, unless, by one means or another, the sauce has been emulsified as a precaution. Furthermore, if a dish contains sautéed vegetables, they will turn brown and dry out when reheated in butter. It is better to use water, in minuscule proportions, to compensate for the loss of the water involved in the initial preparation.

Of course, if a microwave oven is available, the problem of reheating is resolved. What a fine invention!

Why Must Cauliflower Not Be Overcooked?

The various vegetables in the cole family (mustard, brussels sprouts, cauliflower, broccoli, turnips, etc.) contain sulfur compounds, analogous to certain aromatic precursors in onions. In these vegetables, however, the sulfur compounds are bound to sugar molecules and odorless as long as they do not come in contact with an enzyme that transforms them into aromatic compounds. This enzyme is inactive in the acidic conditions of normal vegetable tissues. But when the tissues are broken down, the enzymes come into contact with the odorant precursors, unbinding the sugar molecules and releasing the odorant compounds. The chemical weapon, mustard gas, is synthesized from such derivatives (which belong to the family of isothiocyanates).

The vegetables in the cole family were among the first to be analyzed because their strong, persistent odor when cooked suggested that they contained odorant compounds. Thus, beginning in 1928, it was discovered that the extracts of these vegetables, and their derivatives containing cystine (an amino acid), break down into various odorant compounds, especially dihydrogen sulfide, mercaptan, and methyl sulfide. Finally, these compounds react with one another to form trisulfides.

The longer the vegetables in the cole family cook, the greater the number of these molecules and the more the odor increases. Notably, the quantity of dihydrogen sulfide produced while cooking cauliflower doubles between the fifth and seventh minutes of cooking. The smell soon fills the whole house.

Choose your cooking time according to the degree of tenderness you desire for cauliflower, but do not go too far over that limit!

Why Do Beans Cause Flatulence?

Raffinose, a sugar present, for example, in peas and flat beans, is composed of a chain of three chemical rings, one fructose, one glucose, and one galactose. The table sugar that we eat, composed of glucose and fructose, is broken down by digestive enzymes into its constituent rings, which are metabolized. On the other hand, we have no enzyme capable of metabolizing galactose. It passes intact into the large intestine where it is assimilated by the intestinal flora (especially the bacteria *Escherichia coli*). The microorganisms of this flora release hydrogen, carbon dioxide, and methane. These are the three gases that inflate the stomach and produce those well-known noisy eruptions.

A good way to eliminate galactose from our vegetables is to let them germinate, because this process creates galactosidase, an enzyme that destroys galactose. They can also be soaked, and the water used for soaking and later that used for cooking discarded.

Sauerkraut and the Miracles of Fermentation

We know that sauerkraut is produced by fermenting cabbage in a brine, where the development of certain pathogenic bacteria is blocked while the development of other organisms like *Leuconostoc mesenteroides* and *Lactobacillus plantarum* is encouraged. During this development, the bacteria consume glucose and expel lactic acid, which gives sauerkraut its flavor.

Lactic acid ($C_3H_6O_3$) is half a glucose molecule ($C_6H_{12}O_6$). It is formed through the anaerobic fermentation (in the absence of oxygen) of sugar and glucose, and it is responsible for muscle aches after sustained exercise when the muscles are deprived of oxygen.

Lactic acid is also found in milk when the milk is colonized by bacteria that make use of its sugar, lactose, and break it down, releasing lactic acid. By increasing the acidity of the milk, lactic acid makes it coagulate. That is how yogurt is produced. Similarly, lactic acid is responsible for the characteristic flavor of pickles and other foods preserved in vinegar.

How to make sauerkraut? It is remarkably simple. Shredded cabbage is placed in salt, and water is added to obtain a salinity of about 2.25 percent. At a temperature of 18° to 21°C (64° to 69°F), the bacterium *Leuconostoc mesenteroides* grows and releases, in particular, lactic acid. Then, when the concentration of lactic acid reaches 1 percent, *Leuconostoc mesenteroides* is replaced with *Lactobacillus plantarum*. A good level of acidity is attained after about two-and-a-half weeks.

The Ripening of Tomatoes

For the end of our journey, let us head toward the sun, with an examination of tomatoes, delicious but ephemeral. Initially green, they progress to a ripe state under the sun's heat, becoming juicy and aromatic . . . but not for long, because they soon rot. Half the tomatoes produced, it is estimated, end up spoiling. What a shame!

Could their ripening be controlled, and their rotting avoided? Undoubtedly, because in general with tomatoes, ripening is preceded by increases in the respiration of the vegetable cells and the production of ethylene, a simple organic molecule that acts as a hormone.

T. Oller and his colleagues at the University of Albany have just demonstrated that ethylene is a cause and not an effect of ripening. In other words, one means of slowing down the ripening process in tomatoes consists of putting them in a very well ventilated place, so that they do not remain in contact very long with the ethylene they produce.

Sauces

CREAMY, SATINY, FLAVORFUL

Neither a Juice nor a Purée

Before they sang of Trojan heroes or the adventures of Ulysses, the Greek poets invoked the Muses, who were supposed to ensure the truth of their poetic madness. As modern bard of one of the basic components of cooking—the sauces—I invoke Ali-Bab, that early-twentieth-century French engineer who, upon returning from his numerous world travels, offered gourmands the fruits of his long travel experience. His *Gastronomie pratique* hardly merits its name, but his paragraph on sauces deserves to be quoted:

> Sauces are liquid food combinations, thickened or unthickened, that serve to accompany certain dishes.
>
> Thickened sauces, by far the most important, all consist of a more or less succulent stock, seasoned, and a thickening agent. The number of stocks for sauces is considerable, the number of aromatics very great, and there are many ways to thicken a sauce. Thus, given these circumstances, it is easy to understand that, with the number of possible combinations being infinite so to speak, here lies a veritable gold mine for the treasure seeker.

What does this quotation teach us? That sauces are thickened to various degrees, but that, generally, a sauce is neither a juice nor a purée. With their highly sophisticated consistency, sometimes syrupy, sometimes creamy, always flavorful, sauces must have a certain quality to accompany fish, meat, vegetables, and desserts.

This precept is implicit in Ali-Bab's description. For sauces, the key words are "consistency" and "flavor." If you examine various recipes for sauces that you have already made, you will see that these two basic elements are present every time: a flavorful liquid and a thickening agent.

If the question of flavor has come up in other chapters, the question of consistency, absolutely crucial, has hardly been mentioned until now. Two recent scientific findings, obtained by researchers in Dijon and Nantes respectively, will convince us of its importance.

First, in Dijon, Patrick Etiévant, a physical chemist at INRA, offered tasting panels various strawberry jams in which differing amounts of jelling agents had been added in order to obtain varying degrees of firmness. The same batch of fruit had been used in all of them, and each of the test jams was analyzed for its chemical composition. Verdict: the firmer the jams were, the less flavor they had.

Second, at the INRA station in Nantes, Michel Laroche and René Goutefongea studied liver mousses in which, to make them lighter, they replaced part of the fat with a hydrocolloid, that is, essentially, a starch of water and flour. Once again assisted by taste testers, these two Nantes researchers discovered that the flavor quality of the liver mousses prepared in this way depended on the consistency. The more hydrocolloids the mousses contained, the more they melted in the mouth . . . and the better tasting they were.

Thus not only do we expect a particular consistency from a particular dish, but the perception of flavors and scents depends on that consistency.

A Variable Consistency

These remarkable findings of modern food science encourage us to assemble the proper gear before taking off to explore the great land of sauces.

The notion of viscosity will be useful to us here. We have seen that a sauce is neither a juice nor a purée. Its consistency, or "viscosity," is somewhere in between. This entry point into the matter allows us to imagine how sauces can be spoiled in the hands of cooks who neglect the important principles. They may be too liquid, too solid, too inconsistent, too full of lumps.

Physics has shown modern gourmands that viscosity is a complex subject and therefore more interesting than they might otherwise have imagined. A few simple culinary experiments will clue us in to this new business.

First, let us dissolve sugar in water. As long as the amount of sugar is small, the solution flows like water, but when the syrup is concentrated, it thickens, sticks to the spoon, and flows with more difficulty. The proper equipment will show us that this viscosity, the inverse of fluidity, remains the same regardless of the speed of the flow: a constant shearing stress applied to a simple solution or a syrup generates a constant speed of flow.

For other fluids, such as mayonnaise, béchamel, and béarnaise sauces, the true sauces in short, this law no longer holds. In some cases, the viscosity diminishes when the speed increases; sometimes, on the contrary, the viscosity increases. Thus a béarnaise sauce that seems very thick, nearly solid when it is sitting in the sauceboat, takes on an angelic fluidity when it passes through the mouth, at a speed of some fifty centimeters (about twenty inches) per second. Naturally, the molecular composition of the sauces is responsible for these flow properties.

And at this point let us retreat from our incursions into the territory of pure physics; we have enough gear to set out for the land of sauces.

Is Béarnaise Sauce Warm Mayonnaise?

With regard to mayonnaise, we have previously seen that water, perfectly fluid, and oil, also perfectly fluid, form a viscous mixture, thick enough sometimes to cut with a knife, if they have been combined in an emulsion, that is, into a dispersion of oil droplets stabilized with the help of the surface-active molecules of egg yolk.

The viscosity of emulsions is widely used in cooking. It accounts for the satiny quality of béarnaise sauce, hollandaise sauce, white butter sauce, and even of milk and cream, in which the quantity of fat dispersed in the water can be as high, respectively, as 4 and 38 percent.

Most of the time, sauce emulsions are of the oil-in-water variety. These are dispersions of droplets of a liquid fatty substance into a continuous phase of water. Butter, on the other hand, belongs more to the water-in-oil variety (it is not a true emulsion, however, because part of the fat is solid).

Let us interpret the recipe for hollandaise sauce, of which the famous béarnaise sauce differs only in the seasoning and the quantity of butter dispersed in the aqueous solution.[1]

To make a hollandaise sauce, egg yolks are beaten, all by themselves, so as to mix their constituent parts thoroughly. Then water is added, lemon juice, and salt. The mixture is then heated (in a hot water bath if you are worried that your burner is too hot) and mixed to obtain an initial thickening. At this stage, the egg is forming microscopic aggregates, which add some viscosity to the emulsion. Finally, while whisking, butter is added bit by bit: the whisking separates the fat, which melts into microscopic droplets, and disperses them throughout the mixture, which is, in effect, a water solution. The sauce is removed from the heat as soon as it has thickened and served immediately.

What takes place during these successive operations? First, the surface-active molecules of the egg yolks have been dispersed in the tasty aqueous solution. These molecules are composed of proteins and lecithins.

Then, whisking the sauce while the butter melts separates the fat into droplets, which become coated with the various surface-active molecules already present in the mixture. At the same time, the protein coagulates, forming tiny aggregates that also become dispersed in the aqueous phase. Indeed, both hollandaise and béarnaise are not, strictly speaking, emulsions but rather share the attributes of two physical systems: emulsion and suspension.

Lipids
Water
Surfactants

1 In principle, béarnaise sauce, like hollandaise sauce, is made from diluted vinegar, wine, egg yolk, and butter. In a saucepan, one puts, for example, a half cup of white wine, a third cup of vinegar, a minced shallot, two pinches of chervil, and a quarter pinch of bay leaf. This is reduced until only five spoonfuls of liquid remain, then three egg yolks are added, and the mixture is whisked, away from the heat. Then, over very gentle heat (in a hot water bath), butter is added, bit by bit while beating vigorously, until the sauce thickens. Finally, the sauce is strained to eliminate the minced shallot, and tarragon is added.

Why Does Hollandaise Sauce Thicken?

Why does hollandaise sauce become viscous? Because it is a mixture more complex than pure water, and it flows with difficulty. Remember that it contains microscopic egg protein aggregates and fat droplets, which are bigger than the water molecules and mutually impede one another.

Another effect also takes place. First, the salt adds ions that link to the various electrically charged parts of surface-active molecules. Then, the lemon juice or vinegar causes the protein to become positively charged, which causes forces of electrical repulsion to appear between the egg aggregates and the droplets. All identically charged, the heads of the surface-active molecules repel each other. Their flow is further complicated by this repulsion; the viscosity increases many percentage points. But there is danger: if the temperature is too high, flocculation can occur, and egg protein aggregates can combine into bigger, visible aggregates. Lumps lurk: use your whisk!

Why Is Béarnaise Sauce Opaque?

To make an emulsion—hollandaise, béarnaise, or white butter—we begin with water, which is transparent, and butter, transparent as well when it is melted. Why is the resulting emulsion opaque? Because the light spreading through the sauce is reflected on the surface of the droplets, and it is refracted within the oil. The phenomenon is analogous to what we observe when we put broken glass into a jar: the whole thing looks opaque, even though each individual piece of glass is transparent. Milk's whiteness and the yellow of a béarnaise sauce or mayonnaise result from this same phenomenon.

Why Do Some Emulsified Sauces Fail?

Hollandaise sauce, like béarnaise sauce, walks a thin line. To make it thicken enough, it must be cooked until the sauce almost turns. There are two opposing schools of thought regarding methods for salvaging a sauce that has turned. We will see that the complexity of the scientific problem is equal to the succulence of the sauce.

A hollandaise sauce can fail because the butter droplets melt together (they coalesce) or, worse still, because the aggregates produced from the egg yolk proteins form lumps. Coalescence is reversible, even if it is annoying, but coagulation is more serious. Some cooks claim that you can salvage a turned béarnaise sauce by removing it from the heat, adding vinegar, and whisking it very vigorously (with a mixer, for example). Others maintain that lemon juice works wonders, and still others claim that acidity has nothing to do with it. Adding a little water and whisking is all that is necessary for reprocessing a turned béarnaise sauce. What to believe? What to do in case of a disaster?

Let us think about this. In creating repulsion between the droplets, electrical forces keep them from rising to the surface and melting together, from coalescing. When the béarnaise sauce becomes too hot, however, the droplets move more and more rapidly, and they collide more and more frequently with increasing energy. The energetic barrier between the surface-active molecules is finally broken down, and the droplets coalesce. At high temperatures, the egg proteins coagulate irreversibly and form lumps.

Thus mastering the temperature is crucial. At high temperatures, the droplets collide very often and quickly, which promotes flocculation. Inversely, the difference between the surface tensions of the liquids increases at low temperatures, so the surface-active molecules have more difficulty forming emulsions.

The question, then, is how to strike the right balance and limit the phenomena that will destabilize an emulsion and possibly cause your sauce to turn.

Why Use Very Fresh Eggs?

The freshness of the eggs is important in preparing a béarnaise or hollandaise sauce because the lecithin molecules they contain are better surfactants than cholesterol is; as eggs age, their lecithin is broken down into cholesterol molecules.

In other words, the droplets of melted butter in a béarnaise sauce are better dispersed with fresh eggs than with eggs that are already old.

What Purpose Does Lemon or Vinegar Serve?

Lemon juice in hollandaise sauce and vinegar in béarnaise sauce give them a delicious, slightly acid flavor that perfectly balances the creaminess of the butter. These two acids are not there just for the pleasure of the tastebuds, however. They also ensure that the sauce stabilizes. Why does the acidity of the medium, which produces coagulation in milk, prevent it in béarnaise sauce? Because in the case of milk, the acidity acts on the proteins, whereas in a béarnaise sauce, it acts on different surface-active molecules. These surface-active molecules do not coagulate, and, better still, they retain their surface-active properties in conditions under which proteins coagulate: you can stiffen mayonnaise with a hard-boiled egg yolk!

Moreover, in warm emulsions, the acids break down the intramolecular bonds of the proteins so that the proteins can arrange themselves on the surface of the lipid drops and act as surfactants.

How Can We Salvage a Béarnaise Sauce?

Since béarnaise and hollandaise sauces are emulsions, a leading potential cause for their failure is a lack of water. As with mayonnaise, there must be enough water to accommodate all the droplets of that delicious melted butter that gives these sauces their remarkable satin-smoothness.

Since these sauces are prepared hot, the bit of water present in the sauce at the beginning of the preparation (as itself or in the wine, the egg yolks, the lemon juice or vinegar, or even the butter itself) can become insufficient for two reasons. First, when the proportion of butter becomes significant, it is the water-in-oil type of emulsion that is the most stable. Second, heated water evaporates. Even if you love wine more, do not forget the water!

In addition, if the melted butter droplets coalesce even though your proportions are correct, it may be because you have not whisked the sauce vigorously enough. Do not despair: quickly remove your béarnaise sauce from the heat, let it cool while adding perhaps a spoonful of water to increase slightly the volume of water where fat can disperse and then beat it very hard. You ought to be able to recover the creamy smoothness you lost.

The case of coagulated emulsions is a little more serious, but not desperate. When a sauce is overheated, the egg often coagulates into horrible lumps and the oil almost certainly separates from the aqueous phase. Once again, cool it as quickly as possible, and add a little cold water. Then use your mixer to break up the lumps by agitating the sauce. Sometimes this operation will save you the trouble of making the sauce over again. The proteins will remain coagulated, but the mixer will break them down into tiny invisible lumps . . . except perhaps to the trained taste buds of a great gourmand.

Why Will Vinegar Repair Béarnaise Sauce?

We have seen that salt or acids (like vinegar and lemon juice) increase the solubility of proteins by breaking down some of their intramolecular bonds, improving their emulsifying powers while preventing them from coalescing by creating forces of electrical repulsion. If it is only insufficient mixing that has caused your béarnaise sauce to turn, adding acids and salt will certainly help recover a proper emulsion.

Vinegar may also work more simply, however, just because of the water it contains. Indeed, in certain cases, béarnaise sauce turns because the continuous phase (water) has become too thin. As with mayonnaise, the aqueous phase must be of sufficient quantity to accommodate all the droplets of melted butter. If there is too much butter, the water initially added becomes insufficient, and the oil-in-water emulsion tends to become a water-in-oil emulsion. Unfortunately, this inversion of the emulsion is often accompanied by a separation into two phases.

To avoid this inversion, remember that the egg yolk added to the sauce is only half composed of water; to provide sufficiently for the droplets of melted butter, add a little supplementary water (or vinegar or lemon juice).

When is there danger of the emulsion inverting? It is calculated that spheres all the same size can occupy, at the most, about 74 percent of cubic volume. By this hypothesis, the proportion of oil to the aqueous phase would be 3:1. The spherical droplets of melted butter in a béarnaise sauce are all different sizes, however, and they can change shape, so this ratio can be as high as 95 percent oil and 5 percent water.

The rule here is to consider that the sauce may need water; remember that, as you heat it, the water evaporates.

The Mystery of White Butter Sauce

Some cookbooks recommend, when making a white butter sauce, to reduce a little cream first before whisking in the butter. To understand this advice, let us recall that cream is an emulsion of the oil-in-water type, because there is a higher proportion of water in cream than in butter (which is a water-in-oil emulsion). By beginning with cream, to which butter is added bit by bit while whisking, the desired oil-in-water emulsion is obtained.

Emulsions in the Roast?

Before examining egg and starch as binding agents, let us remember that other sauces are emulsions as well. When you make a roast, for example, the fat drips from the meat into the pan at the same time as the juices, which contain some gelatin with surface-active properties. If you whisk together the fat and the juices (possibly adding a little butter at the end), you will obtain a bound, emulsified sauce.

Often when the roast is a bit overcooked, the water evaporates and only the fat remains. Add a little water or wine to obtain the quantity of water you need to make the continuous phase.

Likewise, when we cook a small beef fillet in a frying pan and deglaze with wine or some other kind of alcohol, we dissolve the caramelized juices in the bottom of the pan. And beyond that, if we want to show off like Molière's Monsieur Jourdain in the kitchen, we can make an emulsion by adding butter or cream.

In these two cases, as for any emulsion, the physical composition is the same: continuous phase, dispersed droplets.

The Mysteries of Meat Glaze

"Gelatin is a surfactant because, dissolved in water, it foams when it is agitated." Thus explained Madeleine Djabourov, physical chemist at the École de

Physique et Chimie de Paris, when I asked her for advice regarding sauces. This remark gave me the key to gastronomy made easy; I pass it on to you.

Many sauces are prepared beginning with a stock: bouquet garni, pieces of bone with some flesh still attached, meat (or fish scraps, for fish stocks), first browned at a high temperature, and then cooked in water for several hours. That is the basic principle. I will pass over the skimming, reduction, and other fundamental but tedious details well covered in the cookbooks. To this stock is added an aromatic base and cream or butter.

What I have gathered from the preparation of these stocks is that they should serve as the base for preparing sauces, because they provide both the flavors and a binding agent. As proof, a stock reduced and placed in the refrigerator will form a colored, gelatinous mass.

Why are flavors and binding agents obtained in preparing a stock? Often we see that the prolonged cooking of fish and fish bones or meat, cartilage, and bones (calf's hoof is famous in this regard) causes the gelatin they contain to pass into solution. Vegetables contribute an indispensable aromatic note.

Since gelatin seemed to me to be the binding agent in sauces based on stocks, I wondered if the stock itself could be bypassed or at least made quickly by adding gelatin to a reduction of fortified meat juices.

This first experiment failed. To obtain enough viscosity, I had to add not one or two grams of gelatin but ten, twenty, thirty. . . . The viscosity was considerable when cold but weak when hot, and the sauce was disgusting.

Why this failure? Madeleine Djabourov's remark enlightened me. If gelatin is a surface-active molecule, then perhaps it is because of its emulsifying properties that it acts to form an emulsion. . . .

In a new experiment, I used only a tiny quantity of gelatin, but I added butter to my sauce, which I whisked. It was a complete success, and my sauce was perfectly bound.

Not content with this success, I decided to take the experiment one step further, because sauces are the gourmand's poison: they make him fat and threaten him . . . with gout and with dieting. Is it possible to retain the succulence of classic sauces without adding all those delicious but harmful fatty substances?

To a certain extent, it is possible. To wine reduced with a few aromatics, I added gelatin and light cream (fat reduced by 15 percent). The latter, unsuitable for preparing sauces under normal conditions (it curdles), proved to be perfect, no doubt because of the large amount of gelatin present.

Binding with Egg

Let us leave the land of emulsions to explore the land of sauces bound with egg. Indeed, we have already approached its border in considering béarnaise and hollandaise. Egg as a binding agent seems to have been discovered around the seventeenth century, but the principle behind it remains mysterious, if the method is simple enough: to a cold or lukewarm aromatic aqueous solution, add egg yolks and whisk them in while heating the mixture; gradually the solution thickens.

Done in this way, the preparation is delicate: if the sauce is not whisked enough or if it is heated too much, that is the end of its lovely viscosity, of the satin-smoothness provided by the eggs. Lumps appear; obviously the proteins in the egg yolk have coagulated.

Good cooks know how to avoid these lumps. By adding a pinch of flour to the mixture, they are able to stabilize the preparation so much that they can bring it to a boil without it turning. I advise the incredulous to try this experiment: using two identical saucepans, pour the same quantity of water or wine and add an egg yolk into each; whisk them identically, heating them in the same way; the only difference between the two sauces will be a pinch of flour, added to one saucepan but not the other.

The results are incontestable. The sauce that contains flour can withstand even boiling without coagulating. The other . . . for the moment let us leave it to its sad, lumpy fate.

What is the effect of this minimal quantity of flour? Apparently the starch in the flour gradually dissolves in the sauce. Its long, very cumbersome molecules seem to prevent the egg yolk proteins from aggregating while at the same time they contribute a viscosity analogous to the one for which they are responsible in a béchamel sauce or in other sauces bound with flour.

Binding with Blood

Similar to sauces bound with egg are those bound with blood. Blood contains many proteins, which, like those in egg, can establish networks that give a thickened texture to sauces.

That is the principle behind *civet*, a game stew prepared with mushrooms, red wine, and blood. The blood thickens the sauce, composed mostly of wine, a little vinegar, and all the aromatics. The same rules apply for binding with blood as for binding with egg: remember the little pinch of flour that makes all the difference!

How Do We Salvage a Turned Sauce Bound with Egg?

A sauce bound with egg turns when the proteins in the egg aggregate into macroscopic lumps instead of dispersing uniformly into microscopic aggregates throughout the sauce. Thus to correct such a disaster, proceed as for a béarnaise sauce in which the eggs have coagulated: a turn with the mixer will break up the lumps and recover the lost satin-smoothness. There is no guarantee, however, that the results will be as fine as if the sauce had been properly prepared. Chefs also filter the sauce through a sieve.

Binding with Starch

Emulsification and binding with egg yolk or blood are not the only means for thickening sauces. Employing a roux or beurre manié thickener is equally efficient . . . on the condition that you exercise good judgment. Misused, flour adds a characteristic, unpleasant flavor or can make for a slightly pasty consistency. All the same, let us be positive: before finding flour's faults, let us examine its benefits.

First of all, how should we use it? Often, a sauce begins with the preparation of a roux. Butter is melted over low heat, then flour is added, and the mixture is cooked for a long time at a temperature just high enough to make it bubble. When the roux becomes golden or deep brown, depending on the sauce, a flavorful liquid is added, and the mixture is slowly heated. The sauce thickens and cooks. Finally, it is skimmed, that is, it is finished by using absorbent paper toweling to soak up the fat that has floated to the surface and by heating it at length to eliminate the solid particles and excess flour.

Why Does Flour Thicken Sauces?

To answer this preliminary question, we must know that flour is composed of proteins and complex sugars that form small granules we call starch. Complex sugars? What do we mean by that? Something very simple. First of all, glucose is a little molecule that serves as fuel for animals as well as plants. Produced by the effect of the digestion of foods, the glucose molecule is circulated through our bodies by the blood. Energy is easily extracted from it by our cells, and, conversely, the molecule is easily regenerated.

Plant seeds especially need energy to develop. Thus plants store glucose in their seeds. Since it is soluble in water, however, glucose on its own would be leached away by the first rain, so it is chained together in long, less soluble molecules, sometimes straight (like amylose), sometimes branched (like amylopectin).

Thanks to weak links between the amylose molecules and the amylopectin molecules, the latter aggregate into small starch granules, between two- to fifty-thousandths of a millimeter in size. In some places, these groupings are orderly and the granules are crystalline. In other places, the granules are amorphous and more fragile.

If starch is useful in making sauces, it is because, heated, the energy of the water molecules is enough to disturb the amorphous areas and establish hydrogen bonds between the starch molecules and the water molecules. Water is gradually introduced into the granules, which swell, forming gels called starch (beginning at 60° to 65°C [140° to 149°F] for wheat flour as amylose molecules leak into the water).

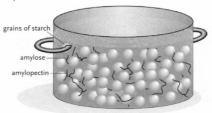

grains of starch
amylose
amylopectin

Why does this thicken the solution? Because the amylose molecules that have passed into the solution are surrounded by water molecules, and also because the swollen starch granules become "microscopically enormous" and cumbersome, making molecular motion difficult. The solution thus becomes viscous. And one final point: that viscosity is maximal when the temperature is kept between 79° and 96°C (174° and 204°F), not quite boiling.

Why Must Roux Be Cooked for a Long Time?

Amylose molecules have only weak thickening power and a floury taste. Thus, to avoid this taste, roux are cooked for a long time in butter before adding the liquid, to break down the amylose molecules into smaller sugars. Flour is an ideal product for the cook because it also contains proteins that react with the sugars through those Maillard reactions I have so frequently mentioned. Not only does cooking the roux eliminate the flour flavor, but in addition it produces odorant and tasty compounds.

If potato starch is used, the long cooking process becomes less important, because the amylose molecules, longer than those in wheat, have a less floury taste. Furthermore, potato starch jells at a lower temperature. It can be used to correct a too-fluid sauce at the last minute.

Why Must a Sauce Bound with Flour Not Be Overheated?

A sauce bound with flour must not be heated at too high a temperature, according to the cookbooks. After mixing the flour and liquid well, the preparation can cook but must not boil. Actually, once the sauce has attained its maximal viscosity, at about 93°C (199°F), it will reliquefy a bit if it boils.

Many conditions favor this reliquefication: heating for a long time after thickening, heating to the boiling point, and too vigorous physical agitation. In all these cases, the swollen granules are broken up into small fragments that flow better than large ones. In addition, a greater quantity of amylose then passes into the solution. A network composed of a greater proportion of amylose is less rigid than one that is principally formed from amylopectin, and the granules are less well maintained.

Why Must Sauces Bound with Flour Remain Liquid While They Are Cooking?

As a sauce bound with flour cools to below 38°C (100°F), the dispersed granules, already further separated by the amylose, begin to form a gel. When the mixture is cooled, the water and starch molecules have less and less energy, and hydrogen bonds begin to hold the molecules more and more securely, eventually reestablishing those bonds that were initially responsible for the cohesion of the granules. The liquid hardens.

This effect should inspire cooks to make their roux-based sauces thinner than optimal for serving at the table. By the time they reach the table, they will inevitably have cooled and thickened.

The Skimming

Skimming is a refined operation. It is the elimination of solid particles in the sauce (see the chapter "Cooking"). Sauces bound with flour, especially, are improved by skimming off some of the solid particles of starch or the lumps formed during the sauce's preparation, as well as the flour proteins, which are not soluble in water. During the sauce's preparation, these proteins coagulate into little solid blocks that must be removed to achieve a perfectly uniform result worthy of Carême and the other great masters of French cooking.

If the famous French cook August Escoffier wanted manufacturers to introduce a gluten-free flour, it was specifically to avoid this long skimming operation. He was not entirely justified, because Maillard reactions, which require proteins, occur during the skimming and also because proteins cooked for a long time are dissociated into sapid amino acids.

In practice, a sauce is skimmed while it is heated, after it has been filtered, in a saucepan tilted in such a way that only one point on the bottom receives heat. Above this point, the sauce is hotter than elsewhere and thus lighter. It rises, and a current is established, with a plume ascending in the center and redescent occurring on the periphery. As this happens, the solid particles follow the current but have the tendency to gather at the center of the saucepan and clump together. They only have to be skimmed periodically to eliminate them.

How Do We Salvage a Sauce that Is Too Thick or Thicken a Sauce that Is Too Thin?

Is the sauce you bound with flour too thick? Beat it vigorously, keeping a close eye on its viscosity. In this way you can break up the swollen granules until the sauce achieves a good consistency.

On the other hand, what to do if your sauce is too fluid? I can recommend beurre manié, a thickener of butter and flour worked together but not cooked. These two ingredients are mixed in equal amounts, and a little ball of the preparation is added to the sauce. The butter keeps the flour from lumping, so that it is gradually released into the sauce. This procedure is only a stopgap measure, however, because it does not avoid that floury taste. Thus it is a good idea to make your beurre manié from corn—rather than wheat—flour.

Why Must Some Fatty Substance Be Used in Sauces Bound with Flour?

Fats do not affect the viscosity of sauces bound with flour, but they affect the impression the sauce makes in the mouth. And, during the preparation of a roux, they coat the flour particles and keep them from lumping in the added liquid. Although their quantity can be limited, it seems difficult to eliminate them entirely.

Why Must Lemon Juice and Vinegar Be Avoided in Sauces with a Roux Base?

If lemon juice or vinegar are heated in the presence of amylose and amylopectin chains, they break down these chains into shorter ones that bind less well with water. The starch granules then gel and disintegrate at lower temperatures. For any given quantity of starch, the final product is less viscous.

A Burning Question

Let Us Eat Well, We Will Die Fat

Why do we like hot pepper, which burns? How can what is good be bad? Before turning to the pepper itself, let me enlarge on the question it poses for us: Is eating harmful?

Brillat-Savarin devoted a few juicy pages to the excesses of dining. Remember, this is his second aphorism: "Animals feed themselves; men eat; but only wise men know the art of eating."[1] And his tenth: "Men who stuff themselves and grow tipsy know neither how to eat nor how to drink."[2]

Very well. So eating or drinking too much is harmful. Today's doctors even try to specify which foods to avoid: certain animal fats, carbon and products of excessive combustion, nitrites used to salt meats . . .

Nevertheless, the danger they see is truly everywhere. Chemists who study Maillard reactions (see page 27, for example), those universal reactions in cooking, find that they produce dangerous compounds of all kinds, and biologists are discovering that amanitoidin, the toxin in the white fungus phalloid amanita, is present in chanterelles and in most other edible mushrooms, though in minute quantities. We must conclude that the harm is in the excess, that it is the dose that makes the poison.

---

1 Jean Anthelme Brillat-Savarin, *The Physiology of Taste; or, Meditations on Transcendental Gastronomy*, trans. M. F. K. Fisher (New York: George Macy Companies, 1949; reprint, New York: Counterpoint, 1999), part 1, p. 3.

2 Ibid., part 1, p. 4.

Does Hot Pepper Burn a Hole in the Stomach?

On the other hand, are the foods that seem harmful really harmful? Pepper, for example: is it as harmful as its effect on the tongue and mouth would lead us to imagine? A burning question that has, finally, just been studied empirically by doctors. David Graham, at the Veterans' Medical Center in Houston, Texas, used endoscopy to observe the effects of hot pepper on the stomach lining of twelve volunteers. He looked for possible inflammations after the absorption of meals peppered in various ways by lovers of highly spiced dishes.

During the first experiment, the volunteers were given a "neutral" meal, consisting of steak and french fries. Then, on another day, they ate the same meal, but seasoned with aspirin (which has a reputation for puncturing the stomach). Finally, on a third occasion, pizza with spicy sausage and various Mexican foods were prepared for them, to which the medical team added as much hot pepper as most people can tolerate.

Endoscopy revealed that aspirin did indeed attack the stomach lining, but that hot pepper had no visible corrosive effect.

The principal pungent ingredient in hot pepper is capsaicin, a phenolic amide $C_{18}H_{22}NO_2$ (or 8-methyl-N-vanniyl-6-neneamide). It stands to reason that this was the first ingredient to be studied. Its effect on the intestinal wall was compared to that of aspirin. Capsaicin had no visible effect. Neither did crushed red pepper deposited directly into the intestine with the help of a cannula.

On the other hand, Tabasco sauce deposited directly into the stomach produced an inflammation of the stomach lining. Why? Because it contains acetic acid; vinegar is a solution of acetic acid in water. In fact, the concentration of acetic acid in Tabasco is two times higher than the concentration of acetic acid in ordinary vinegar.

Thus, if red peppers stimulate nerve endings that register pain, especially in the mouth, they do not have an actual corrosive effect. They stimulate salivation, activate digestion, cause burning sensations in the anus, and provide a feeling of well-being after a meal. Why? Perhaps because they stimulate the release of endogenous opioid substances, similar to morphine in their effect on the pain-sensitive nervous system.

So let us no longer be afraid to use hot pepper. Its fire will not consume us.

The Salad

AN OASIS OF FRESHNESS

Should the Salad Be Prepared in Advance?

Salad, with the vinaigrette that accompanies it, is a dish that the gourmand has never managed to fall in love with completely. It is a delicate, refreshing, and welcome complement to a big meal, but it "kills" the wine because of its acidity. If you serve salad, give your guests only water to drink with it, and change mounts, as riders say, for the cheese and dessert courses. The salad interlude requires that a completely different wine follow the one you served with the meat.

How should a salad be prepared? We all think we know how: you wash the salad, add a vinaigrette dressing, and toss it.

Not so fast! Did you know that if you are serving a mixture of many different salad greens, you would be well advised to toss the toughest varieties first and then add the tender varieties? Did you know that the vinaigrette should not be added before you are ready to toss the salad? Did you know that vinaigrette is a different dressing depending on if there is more or less oil?

Vinaigrette

Let us begin by examining the composition of the vinaigrette. We have seen with regard to mayonnaise that we achieve a mixture of oil and water by forming an emulsion, that is to say, a dispersion of oil droplets in water or, conversely, a dispersion of water droplets in oil. Composed of vinegar, oil, salt, pepper, and mustard to taste, a vinaigrette is just such an emulsion. The vinegar is a solution of acetic acid in water; the oil is . . . the oil.

Normally, oil does not dissolve in water. It is only when a mixture of oil and water is vigorously agitated that the oil droplets achieve a suspension in the water. And this is only temporary, moreover, because these droplets, which are lighter than water, rise again to the surface, merge, and reform a separate oily phase.

Nevertheless, if the oil droplets are small enough, their separation is slowed, because their dispersion hampers the rising process. In addition, the mustard in vinaigrettes increases the stability of the emulsion: whisked into the vinaigrette at the same time as the oil droplets, its surface-active molecules bind to the oil molecules along their hydrophobic extremity and to the continuous, aqueous phase along their hydrophilic extremity. They thus form a link between the oil and the water.

This description only applies as long as the proportion of oil is not too great. When the water and oil are present in equal quantities, the oil forms droplets that disperse in the water because it has the tendency to form droplets. On the other hand, if the proportion of oil is increased, it is the water that will be dispersed in the form of droplets in the oil. In the presence of mustard, this transition takes place when the proportions exceed two parts of oil to one part of water.

In any case, these emulsions are more transient than a mayonnaise emulsion. Left to itself for a little while, a vinaigrette separates into vinegar at the bottom and oil on top.

The Seasoning

Finally, a word regarding the use of this water-in-oil (or oil-in-water) emulsion for salads.

Oil adheres better to the surface of vegetables than water does, but both substances do harm to the color. They penetrate the surface thanks to fissures in the waxy cuticle that coats vegetable leaves, like salad greens, and they drive out the air that, by refracting the light, gives the leaves their beautiful green color.

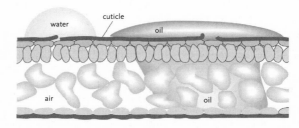

Whether your vinaigrette is an oil-in-water emulsion or a water-in-oil emulsion, toss it with your greens only at the last minute if you want to serve a very green salad.

Yogurt and Cheese

Acid and Rennet

I warned you in my introduction, my dear guest of gastronomical literature, that I would only lead you where your own cooking resources would suffice. So do not take me to task for offering too little advice when it comes to cheeses. These appear at the table in just the state in which they were acquired. At best, through a little elementary care, you might continue the maturing process accomplished by your cheese maker.

Nevertheless, the good gourmand will be tempted to make cheese and will need some information. It is important to know that cheese is obtained through the coagulation of milk. I will only add that milk coagulates because the micelles of casein (the protein that represents 85 percent of the proteins in milk, which constitute 4 percent of the total milk) aggregate when conditions lend themselves to it.

In addition, the gourmand should know that the fatty substances in milk are formed during the lactation of the cow, sheep, or goat. The mammary cells have a surface that forms a protuberance from which are released fat globules about 2.5 micrometers in diameter. The interior contains fats but also vitamin A and cholesterol. The outer membrane contains surface-active molecules, which ensure the fat emulsification.

When milk is made to curdle, either by adding rennet (extracted from the fourth stomach of calves), salt, or acid, the casein coagulates, while the curds retain some of the fats and a few proteins in solution. The effect is the same as the one that allows us to fry an egg sunny-side up successfully (see page 46). In the presence of ions provided by the acid or salt, the casein molecules no longer exert the electrical forces on one another that are responsible for their repulsion, and the casein micelles aggregate.

Have you already done the experiment of salting or acidifying hot milk?

Careful Attention

Coagulating the milk is only the first step in making cheese. It is followed by a draining process, a salting process, and then a maturing process, which takes place with the help of selected microorganisms.

During the draining process, manufacturers use lactic bacteria with the rennet, which acidify the environment (by releasing lactic acid). The salting takes place through immersion in a brine. Then comes a sprinkling of microorganisms to start the maturing process, which gives each cheese its unique character.

Why Does Cheese Smell?

Cheese smells because a considerable share of the fatty acids is in a free form (that is, not incorporated in the triacylglycerols) because of the lipase enzymes in the microorganisms used for the maturing process. The *Penicillium camemberti* microorganism, for example, has an important function in attacking and transforming the fats. Localized especially near the rind, the bacteria hydrolyzes the triacylglycerols, weakens the periphery of the cheese (in well-ripened camembert), and releases the gas we call ammonia. It is this smell that repulses some, depriving them of the immense pleasure of tasting, with well-ripened Camembert cheeses, the flavor of France.

The presence of ammonia around cheese in the process of maturing seems to contribute to its favorable evolution. A word of advice, if you have acquired some insufficiently aged Camembert, put it in a tightly closed bag in a high place in your kitchen.

Preparing Yogurt

How does yogurt form? The recipe is simple: place a spoonful of yogurt in a pot full of warm milk and heat slowly for a long time (many hours)—for example, in a double boiler or in the oven. The milk forms into a mass. It is a kind of multiplication of the little yogurts.

The principle molecule in this process is lactic acid, which can be considered as half a glucose molecule, the fuel for our bodies. Lactic acid forms through the fermentation of glucose and other sugars in the absence of oxygen.

Milk, which contains sugars, is rapidly colonized by the bacteria that act on the milk sugar, lactose, and break it down, releasing lactic acid. The lactic acid coagulates the milk according to the same phenomenon used in making cheese. The coagulation of casein produces a gel that traps the water and fat droplets.

Fruits of the Harvest

Why Do Apples Turn Brown
When They Are Cut?

When an apple is cut or peeled, its surface, which is initially white, turns brown within minutes. Apricots, pears, cherries, and peaches do not brown, but, even worse, they blacken! Bananas and potatoes turn pink before turning brown. Lemons and oranges, on the other hand, do not brown. Does their natural acidity protect them?

Absolutely. If certain fruits brown when they are cut, it is because the knife damages some of their cells, releasing their contents and especially some enzymes that were enclosed in special compartments.

More precisely, the enzymes, called polyphenolases, oxidize the colorless polyphenol molecules of the fruits into orthoquinone compounds, which are rearranged, undergoing oxidation and polymerizing into colored molecules that are cousins of melanin (melanin is the molecule that makes us turn a beautiful bronze color when we have been exposed to the sun).

Acidity slows these reactions, because it limits the action of the enzymes. In addition, the ascorbic acid in lemons and other fruits of the same family (oranges, grapefruits, etc.) is an antioxidant. These are two reasons it is a good idea to squeeze lemon juice over cut fruit if you want to retain its original color. You can also use pure ascorbic acid (vitamin C) from the drugstore if you want to avoid the taste of lemon.

How Much Sugar for the Syrup for Preserved Fruit?

Those who understand the physical phenomenon of osmosis, already discussed with regard to braising, can succeed in preserving fruit in syrup. The word "osmosis" is a mouthful, but the phenomenon is simple. In a liquid, a drop of ink disperses gradually so as to occupy all the liquid; its concentration is equalized. There is nothing mysterious about that. The molecules in a liquid move incessantly, so that, through chance collisions, the molecules of the drop are distributed throughout the liquid.

In fruits preserved in syrup, the identical phenomenon is at work. When fruit is cooked in plain water, the sugar from the fruit will tend to pass into the water in order to equalize the concentration of sugar, and the water in the external environment will pass into the fruit cells to dilute the sugar found there. As sugars are large molecules, only the water moves, so the fruit swells with water and then explodes.

On the other hand, if fruit is cooked in a solution in which the concentration of sugar is higher than the concentration found in the fruit, the water in the fruit tends to be released from the plant cells in order to lower the concentration of sugar in the solution. The fruit shrivels. Consequently, cooking the fruit in a sugar solution equal in concentration to the fruit's will best preserve the fruit's natural appearance.

The same phenomenon takes place in preparing marrons glacés. First, the chestnuts are cooked for a long time in water in order to soften them thoroughly. Then they are peeled, and, when they have cooled, slowly (so as not to break them) immersed in increasingly concentrated syrup (flavored with vanilla). Gradually, the sugar saturates the chestnuts.

Ices and Sorbets

How Must Ices Be Agitated?

The scourge of ices and sorbets is the ice crystal. When it seems to be absent, the dessert is delicious, velvety smooth, and melting in the mouth. But when it is present, a horrible sensation of broken glass in the mouth ruins the pleasure that the dish's thousand shimmering reflections had promised.

Physicists and especially crystallographers are well acquainted with crystals. They know that in order to obtain large ones, the parent solution must not be moved and as slow a growth as possible must be encouraged.

The cook, who desires the opposite effect, must thus agitate such solutions as much as possible in order to prevent large ice crystals from forming. At the same time, the cook wants to introduce air bubbles in order to obtain a light consistency.

How and when should this be done? At the beginning, agitation serves no purpose: the preparation must first cool. As long as the temperature is above 0°C (32°F), no ice crystals can form. Furthermore, introducing air bubbles at this stage is useless because the preparation is still too liquid to retain them. And, finally, the cream that is present in the preparation is in danger of turning to butter if it is agitated too much.

But once you round the Cape of Zero, heave-ho!

Must the Preparation Be Placed in the Freezer Hot or Cold?

What a question! Common sense tells us that the freezer will work more efficiently if the preparation is already cold. Common sense is not always right, however. Hot water freezes more quickly than cold water.

This effect was studied by Ernesto Mpemba in Tanzania, who prepared a recipe that called for heating the milk, incorporating the sugar, letting the mixture cool to room temperature, and freezing it. One day, he forgot to let the mixture cool and discovered that his preparation froze much more quickly than when he faithfully followed the recipe. He published the results of his subsequent studies in the *Journal for Physical Education*.

Why does hot water freeze more quickly than cold water? It has been claimed that it warms the receptacle in which it is deposited, which melts the ice, and thus a better thermal contact is then established with the freezer. This explanation is insufficient, because the effect occurs even if the receptacle is insulated with little wooden chocks or polystyrene foam, for example.

In fact, three different effects seem to come into play. First, convection: that is, the movement of liquid when its temperature at the top is not the same as at the bottom; the difference in density produces flows that homogenize the solution. Second, cold water dissolves more gas than hot water; with the gases removed, hot water cools more quickly. Third, a hot solution loses water through evaporation, so that, in the end, there is less water to cool.

Instant Ice Cream

In 1901, at the Royal Institution of London, Agnes B. Marshall invented an ideal method for preparing ice cream or sorbet. It is ideal because, using her process, the ice crystals are tiny, as desired, and the preparation is extremely light because of the countless air bubbles introduced into it. And last but not least, the preparation can be made at the table, before your guests, in a few seconds. What is this marvelous contribution to gastronomy?

Agnes Marshall proposed abandoning the classic, old-fashioned ice cream maker for liquid air, or, more precisely, liquid nitrogen. This transparent liquid, present in all chemistry and physics laboratories, is nothing other than nitrogen from air that has been cooled to –196°C (–320°F). I do not have to tell you that that is very cold.

When it is (slowly) poured into a preparation for ice cream or sorbet, it vaporizes immediately, absorbing the preparation's heat and instantly freezing it. Penetrated by the cold, the preparation becomes filled with tiny ice crystals, while the liquid air passes into a gaseous state; the air bubbles are trapped in the ice cream or sorbet.

The whole thing takes place in an impressive cloud of white mist, the same kind that is used in shooting films when the director asks for fog. A guaranteed success![1]

---

1   Be careful when handling liquid nitrogen: wear hand and eye protection and open windows to avoid filling the room with nitrogen gas.

Cakes

LIGHT AND MELTING

A Base Both Robust and Light

To become a good cook, Escoffier said, you must first try your hand at pastries, because that is the best school for learning correct proportions. Let us add that pastries are also a wonderful domain for the physical chemist . . . and for the gourmand. Isn't that where we find whipped cream, mousses, candied fruit, and a thousand other preparations that science can help us make successfully without mistakes?

Many cakes begin with a solid base that supports the rest of the creation. How to obtain one that is airy and melting? Spongy or foamy textures are essential. The walls of the bubbles or cells, like the walls of a honeycomb, provide a kind of strength that modern engineers have learned to use in their work. A structure full of bubbles retains a tenderness that harmonizes with the cream, often whipped, that it supports.

Nevertheless, the foam of beaten egg white is too fragile to support a whole cake, and the foam in a soufflé has the disadvantage of collapsing after being cooked. What to do? It must be reinforced.

Let us examine two types of reinforcement analyzed by Peter Barham, my friend from Bristol, whom I mentioned earlier. The first, used in meringues, rigidifies the walls of the bubbles in the foam. The second, borrowed from the building industry, adds a load (an edible one, of course): flour or sugar.

Meringue Foam

When plain water is beaten, a few bubbles form and then collapse. On the contrary, when an egg white (which contains 90 percent water) is beaten, an excellent foam is obtained, stable for many hours. The reason? "Surfactants" is the key word here, as we have seen many times.

Egg white foam, as we have already seen with regard to soufflés (see pages 50–51), is formed by trapping air bubbles in a liquid. In plain water, air bubbles rise to the surface; beating egg whites, however, produces the forces that will stabilize them. We benefit from the presence of surface-active molecules in the egg, that is, molecules having a hydrophilic part (which binds to the water molecules) and a hydrophobic part (which resists being in the water and thus positions itself instead in the air).

Small bubbles are more sensitive to the surface forces and to the forces provided by the surface-active molecules. They will thus form a more stable foam and better withstand any pressure applied to them because this pressure will be distributed over a greater number of bubbles.

To make a meringue, powdered sugar, which dissolves easily in the bubble walls, must be added. This is the classic recipe: two tablespoons of sugar per egg white for a soft meringue and four tablespoons for a firm meringue. The mixture is dried in a warm oven, and when some of the water has disappeared, only the rigid structure remains, composed of surface-active molecules, sugar, and water, which is bound to this structure and will evaporate only after prolonged heating, which, naturally, must be avoided.

In the oven, the heat expands the bubbles and vaporizes the water, which makes the meringue swell. Simultaneously, the heat from the oven coagulates the various egg white proteins, which permanently rigidifies the formed bubbles. Ideal baking will create a firm crust with a soft, supple interior. Bake for forty minutes at 120°C (248°F) first and then for two hours more at 100°C (212°F) for meringues with a soft inside, or one hour more at 110°C (230°F) for harder meringues.

When Should the Sugar Be Added?

When should the sugar be added to the egg whites when preparing a meringue? Before beating them or after?

All cooks are sure about this: the sugar must only be added when firm peaks have already formed. Why? Because it dehydrates the proteins, especially if it is very fine and, as in the form of a sugar glaze, very well dispersed. The physical phenomenon is, again, a matter of diffusion. If sugar, which contains no water, comes into contact with proteins, with properties dependent on their bonds with water, the water tends to leave the proteins to dissolve the sugar. If the sugar is added too early, the foam cannot form, and the whites whip up poorly.

Opening the Oven Door

For a meringue as for a soufflé, do not open the oven door while it is baking. The bubbles of air and vapor, which swell as they heat, are in danger of rapidly deflating. If this happens, as the meringue continues to bake, the egg whites solidify before they have time to reinflate. Instead, use an oven with a glass door and prepare to be patient.

The Scylla and Charybdis of Meringues: Overbeating and Egg Yolk

Be careful not to beat too much. If you denature the proteins too rapidly, air will be introduced in insufficient quantities at the time when the bonds between the proteins should be established. And if you continue beating after these bonds are established, the number of bonds between the molecules will continue to increase in the foam, which will expel the water that is normally bonded to the molecules. You will see it beading on the surface.

Now that you know the Scylla of meringue, let us examine the Charybdis. It is frequently said, mistakenly, that it is impossible to obtain a good foam if you beat the whites with even a trace of yolk in them. It is true that it is much more difficult to obtain a stable foam when yolk is present, but it is not impossible.

We know that the cholesterol in egg yolk is a molecule that tends to bond to the hydrophobic groups of the denatured proteins and thus prevent those groups from participating in the formation of the foam. Because of this, when yolk is present in the whites, a greater quantity of proteins must be denatured to bond with the cholesterol, and beating takes much longer than with egg whites alone. Furthermore, the speed of beating required to denature a molecule increases as the viscosity diminishes (stress is proportional to viscosity); thus adding a pinch of sugar or salt, as recommended by cooks, makes the beating easier by increasing the viscosity.

A Soft Base

While some cakes contain meringue, others have a spongy base. The principle is analogous to that of meringues, but instead of baking stiffly beaten egg whites mixed with sugar in order to obtain a rigid surface structure, a softer texture is retained by adding a load: flour.

As with the sugar in meringue, the flour is only added when the whites are very firm; otherwise, the fine starch particles in the flour will capture the air bubbles and make the foam collapse. Thus the preparation is mixed like a soufflé, by folding the foam over the flour with the help of a spatula that is manipulated as though one were cutting a tart. As soon as the color is uniform, mixing stops.

Likewise, a fatty substance is added, generally melted butter, to provide a silky texture and slow down the recrystallization of the starch. As before, the foam is folded over the warm, melted butter, because the fat molecules in the butter tend to bond to the hydrophobic groups of proteins and make the foam collapse.

Finally, to prevent the foam prepared in this way from collapsing, it is baked. This increases the denaturation of the proteins and leads to the formation of permanent intermolecular bonds, transforming the semiliquid mixture into a rigid sponge.

In the oven, many simultaneous reactions harden the interface of the bubbles and thus allow them to withstand the pressure caused by the expansion of the air and the formation of vapor. In practice, the speed of hardening and the formation of vapor are not equal, which means that the bubbles grow and the volume of the cake increases by 10 percent. A good way to test if a cake is baked is to insert a knife in the foam and see if it sticks to it. If the foam sticks to the knife, the cake has not finished baking.

When the cake is taken out of the oven, it cools, the gas in the bubbles contracts, and the vapor condenses, which reduces the internal pressure and causes the cake to collapse if it is not completely baked.

It is quite easy to avoid this inconvenience. You can deliberately make some of the bubbles burst by dropping the cake, in its pan, from a height of about ten centimeters (four inches). The result will be less beautiful than what comes out of the oven, but you will not have the displeasure of watching the base collapse unevenly under the weight of your various garnishes.

Whipped Cream

Finally, the ultimate step remains: preparing the garnishes, often composed of a mixture of whipped cream and dark or red fruits. Whipped cream is a type of foam, once again, but its stabilization is the result of a different effect. Actually, milk and natural cream are composed of small globules of fat, and their suspension in water is stabilized by surface-active molecules, such as casein.

In milk, the proportion of fatty substances is about 7 percent. In thin cream, it reaches 18 percent, and in thick cream, it rises to 47 percent. In butter, on the other hand, the proportion of fatty substances is 83 percent, but the emulsion is reversed. Here droplets of water are dispersed in the fat.

In whipped cream, we are looking for a dispersion of air and water in the fatty substance. This inversion of the emulsion that constitutes cream is obtained by beating it.

The same stabilization phenomena are at work in whipped cream as in stiffly beaten egg whites. Viscosity is important for stabilizing the foam. Thus, as thick a cream as possible should be used, and the ingredients should be refrigerated before beating them to increase their viscosity further. To avoid the formation of butter, beating can be done in a bowl that is sitting on crushed ice.

When the cream is stiff enough to support its own weight and also that of the garnishes, that is, when the fat globules have been divided enough to coat the air bubbles and stabilize them, alternate layers of sponge cake, meringue, fruit, and whipped cream will be piled on top of one another. A drop of alcohol will charm adults and give children a taste of good things to come. And a dusting of chocolate, for example, will add the finishing touch to your creation.

Pastry Dough

TART, SHORTBREAD, AND PUFF PASTRY

Why Must Dough Be Allowed to Rest Before Baking It?

Everyone knows that pastry doughs are basically flour, water, and butter. Nevertheless, tart dough is nothing at all like puff pastry, which differs considerably from sweet shortbread dough. Why do the same ingredients produce such different results? Because the hand of the pastry cook comes into play. Let us examine how.

The simplest dough is prepared with just flour and water, mixed in proportions that yield a substance with the consistency of thick putty that does not stick to the fingers. In the making of this dough, water is introduced between the countless starch granules in the flour, and it binds them into a coherent mass, linking to proteins (we shall later explore "gluten"; please be patient). In effect, as soon as water comes in contact with flour, it penetrates between the granules through capillarity.[1] If one rolls out this dough into a round two or three millimeters thick (about one-sixteenth of an inch), one will have an unleavened flat bread.

While it bakes, the temperature of the water increases, and the granules inflate and form starch that gradually binds together as the water evaporates. A single mass of hard dough is formed.

---

1   To see capillarity in action, make a *canard*, a sugar cube dipped in coffee. When you immerse the base of the sugar cube in the coffee, you can see the coffee rising in the sugar as it is drawn up by the forces of capillarity between the grains.

Pastry dough differs from this because it contains butter (or margarine, or some other fatty substance) that separates the particles of flour from one another and thus helps the dough retain a certain suppleness. As before, mixing the flour and the water forms a starch, but the fat separates the individual starches. After baking, the dough is still crumbly because the starch granules have remained fairly separate. The cohesion is the result of the butter, which, in cooling, forms a sort of smooth cement. The proof of this? Warm pastry dough is more crumbly than the same pastry dough when cool. Thus it is better to take cakes and tarts with a pastry crust out of their pans only after they have cooled.

The preceding examination of these two simple doughs teaches us an important culinary lesson. Since the goal is to obtain the jellification of starch, the preparation of the dough must not be rushed. The water must have time to migrate between the granules and then to penetrate them in order to make them swell. That is also why recipes tell us to let the dough rest before baking it.

Limited Kneading

Good cookbooks warn against kneading pastry dough too much. The reason behind this advice? Because flour is not composed only of starch. It also contains proteins, some of which form what is called gluten.

The term "gluten" was used as early as 1742 by the Italian chemist Giacomo Bartolomeo Beccari (1682–1766). Wanting to investigate the composition of flour, he kneaded flour with water to make a dough and then kneaded that dough in water. The white powder that escaped with the water was starch; the elastic yellow mass that remained was gluten, made of some of the proteins of flour.

Gluten proteins are thirsty for water and in its presence form a very tough, though elastic, network. Prolonged kneading, which makes the gluten proteins coagulate, produces a very tough dough.

It is worth noting that when the dough is being used to make brioche, it is better first to mix the butter with the flour in order to coat the particles of flour with the fatty substance. Water added subsequently will form the necessary starches, and, if the kneading is limited, it will not produce the tough elastic network of gluten that would inhibit the brioche from rising.

A Thousand Layers . . .

Now, puff pastry is "exponential." If you fold one layer of dough once into three (one turn), you form an ensemble of three stacked layers. If you fold this into three, you get nine layers, and if you make six successive turns, always separating the layers, you get 3 × 3 × 3 × 3 × 3 × 3 layers, or 729 altogether! In French, this is called *pâte feuilletée*, foliated pastry, a name it certainly deserves. The butter keeps the layers separate, and that is how they can bake without merging together. In addition, the folding of the dough traps air, which expands in baking and helps to separate the layers further, as does the vapor that forms from the water present in the dough. The pastry becomes lighter as it expands in baking.

Having established these principles, I would like to perform a public service by offering the recipe for puff pastry here. Let me explain. Having obtained very different results with the various recipes given in cookbooks, I methodically compared them and finally arrived at the following.

Make a dough by mixing flour and water. How much of each? For 250 grams (8.8 ounces, or a bit more than a cup) of flour, use a maximum of 1.5 deciliters (slightly more than 5 ounces, or a little less than two-thirds of a cup) of water. The proportion of water can vary considerably according to the flour, which can contain more or less protein, and also to temperature. When you have obtained a uniform soft ball of dough, knead 250 grams of butter until it is the consistency of the dough. Roll the dough into a fairly thick square, about 20 centimeters (about 8 inches) on each side, and on top of it deposit the butter in a square about 10 centimeters (about 4 inches) on each side, in such a way that the corners of the butter square come to the centers of the sides of the dough square. Fold the four corners of the dough over the butter to form an envelope around it. Then roll the dough in just one direction to obtain a rectangle, which you will fold in three. Having once again formed a square, you will turn it one-quarter turn, and you will roll it once more into a rectangle, which you will once more fold in thirds.

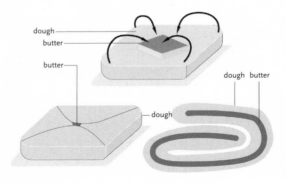

Let the dough rest in a cool place for twenty minutes, and then repeat the operation of rolling, folding in three, turning a quarter turn, and folding again in three. Return the dough to a cool place, and, before baking, repeat the operation of rolling, folding in thirds, rotating a quarter turn, rolling, and folding in three. Bake for about 40 minutes at 180°C (392°F), after trimming the edges and making light incisions on the surface brushed with an egg yolk mixed with milk. Your puff pastry will be superb.

Sweet Dough

The pastry doughs I have examined consist of flour, water, and butter. Sweet pastry dough, used for a dry shortbread pastry that breaks and crumbles in the mouth, is again obtained by mixing flour and butter, but added to these ingredients are sugar and egg yolks. The dough, made without water, is crumbly and difficult to roll out; to be successful, do not apply too much pressure on the rolling pin. Then bake.

During the preparation of the dough, the butter and egg yolk penetrate between the granules of starch and sugar. However, since sugar captures more water than starch, a dispersion of starch granules becomes dispersed in a syrup, preventing gluten formation. The (weak) cohesion in sweet pastry doughs is due to the egg yolks, which coagulate into a network that traps the various granules.

The Bubbles in Sponge Cake

Sponge cake dough does not contain any water, either. Its lightness, once again, is due to the absence of jellified starch in the uncooked dough. To make it, egg yolks and castor sugar (a fine-grain sugar somewhere between granulated and powdered sugar) are whisked together, incorporating as much air as possible. The egg yolk forces itself between the sugar granules, which, in the fatty medium, remain intact and separated by the millions of air bubbles.

Then, using a spatula, stiffly beaten egg whites are added in order to form a kind of light sponge. Finally, add just barely melted butter and mix it in gently with a fork to obtain a viscous batter that does not easily run. Pour the batter into an ungreased cake pan with high, smooth sides and place in a very hot oven. During the baking process, all the bubbles will expand, vapor bubbles will appear, and the surface will take on color through various chemical processes, caramelization in particular. The sponge cake rises and turns golden, and in 15 to 20 minutes, the blade of a knife that pierces it will emerge clean and dry, a sign that it is fully baked. It is left to cool in the pan, and the result is an airy, delicious cake.

Why this mode of operation and these results? First of all, the egg whites provide proteins that coagulate during the baking, stabilizing the network of inflated bubbles, while the sugar, in melting, reacts with the egg yolks. The preparation is similar to a soufflé, but the sugar serves as a load, stabilizing the network. In addition, the various reactions involved in caramelizing the sugar and cooking the eggs give the sponge cake its pleasant taste.

Finally, a note on génoise. This sponge cake differs from the one described above in that the egg whites and yolks (as well as the rest of the preparation) are whisked together. How do the egg whites rise into stiff peaks with the yolks present? Isn't this much too difficult? Difficult, but not impossible. You must whip them for a long time (up to about a quarter of an hour, with a brief rest in a hot water bath), remaining intent on introducing air into the mixture.

Leavened Doughs

Without addressing bread, which merits a chapter all its own, let me examine the question of leavened doughs in pastry making. For these doughs, we often use what are incorrectly called in French "chemical yeasts."

What are we talking about? First of all, let me make it clear these are not biological agents like yeast, single-cell microorganisms that release carbon dioxide when they are in the presence of sugar. "Chemical yeasts," also known as leavening powders, are excellent leavening agents for the finest cakes. They are compounds that release carbon dioxide when they are in the presence of heated water. Sugar is not required for their action.

Often these leavening powders are a mixture of sodium bicarbonate, an acid (or, often, two acids, such as tartaric acid and sulfate acid of sodium and aluminum), and starch, which acts as an excipient, separates the particles of acid and bicarbonate, and prevents the active components from reacting prematurely.

Leavening powders act twice: once, at room temperature, because of the action of the tartaric acid on the bicarbonate, which releases carbon dioxide and produces little bubbles in the batter; then a second time, because of the action of the aluminum salt at high temperatures, which increases the size of the bubbles and makes the dough lighter.

This has nothing to do, however, with the live leavening agents, yeasts, like the baking yeast, *Sacchoromyces cerevisae*, that die when the temperature becomes too high. On the other hand, baking yeast is much more effective for leavening a dry dough, like bread dough, as opposed to a batter.

Sugar

What Happens When Sugar Is Heated?

Louis XIII's cook, Jean de la Varenne, said that "a man who attaches great importance to dessert after a good meal is a fool who spoils his spirit with his stomach." Some gourmands will share his opinion, but many of us, even as adults, have not lost our immoderate taste for sugar and its various forms . . . like caramel.

Why its golden color? Why its inimitable flavor? Table sugar is composed of a molecule called sucrose, a glucose ring with six carbon atoms, bound by an oxygen atom to a fructose ring.

When this molecule is heated, it undergoes a complex series of decompositions, and, since each molecule possesses many oxygen atoms, rearrangements are possible. The molecules break up, and little volatile fragments, such as hydroxymethylfurfural, either evaporate or dissolve in the substance and give it its odor, as fructose dianhydrides form, react with simple sugars, and make the caramel mass.

Why Are We Advised Against Using Aspartame in Cooking?

Abhorred by those who aspire, as Brillat-Savarin said, "to be unfinished creatures,"[1] sugar is sometimes deposed by various sweeteners, such as aspartame.

---

1   Jean Anthelme Brillat-Savarin, "Meditation 23," sec. 114 in *The Physiology of Taste; or, Meditations on Transcendental Gastronomy*, trans. M. F. K. Fisher (New York: George Macy Companies, 1949; reprint, New York: Counterpoint, 1999), part 1, p. 262.

Why are we advised against using the latter in cooking? Because its molecule is composed of a molecule of aspartic acid bound to a molecule of phenylalanine. When heated, the two parts separate, and their two tastes replace the one sweet taste of aspartame. Aspartic acid may be tasteless, but phenylalanine is bitter.

We should also remember that aspartame slowly breaks down in aqueous solutions. Drinks sweetened with aspartame should be used promptly; if stored, they will become bitter.

Bread

How Do We Make Good Bread?

Many city dwellers have forgotten the taste of their origins. Sometimes they have also forgotten what, after a millennium of civilization, has become second nature to humans: the preparation of bread. A few elevator stops separate them from the baker, who lets them benefit from his professional expertise, his specialized equipment . . . and his appetizing breads.

Why should we go to the trouble of arduous kneading and baking, with sometimes mediocre results? Will we ever attain those golden baguettes, those crusty, sweet-smelling breads neatly lined up behind our pleasant baker? Where can we find the flour and yeast we would need to put our own daily bread on our table? Where do we find the time to make bread? And, finally, what about the techniques necessary to bring about the miracle of bread making?

There are many among us who have never made bread and hesitate to give it a try, for fear of failing. But if they only knew! If they only resolved to make their first loaf, the only one that counts! They would be proud of work well done and feasting on products much superior to the breads that, more and more frequently, bakeries that are only fronts do not even make themselves but buy already prepared.

I invite you now to a celebration of domestic bread making. I want to help you share in this child's play, the preparation of a good homemade bread. And to help you avoid witnessing your dreams of a baguette reduced to the disappointment of a poorly baked doughy lump, this is my proposal: that you discover the physical chemistry of bread by putting your hands in the dough.

From Water, Flour, Yeast . . .

If our earliest ancestors who made bread obtained results that encouraged them to continue experimenting, it is because the principle behind bread making is simple. To make bread, we need only water, flour, two hands to knead, and an oven. Three operations will do it: kneading, fermenting, and baking. Nothing is simpler, nothing more ancient, than the techniques of bread making. They have hardly changed since they were practiced in Egypt three thousand years ago.

First let us assemble the ingredients: 12 deciliters (about 5 cups) of wheat flour, 5 deciliters (about 2 cups) of water, 25 grams (a bit less than an ounce) of yeast, and a teaspoon of cooking salt. In a large bowl, we dilute the yeast in the water; then we add the flour and salt. We mix, knead, and pummel for a long time, and then we form a ball that we pat with flour. We cover the bowl with a cloth and let the dough rest for three hours. This is the first step. The dough swells up.

We then knead it until it returns to its initial volume. Then we form it into a sausage shape that we lay in a buttered baking pan and cover with a cloth. The dough rests for another three hours in the kitchen or overnight in a cool spot. During this phase of preparation, the dough must rise well and overflow the pan. Be careful not to touch it! After making small incisions in the top of the loaf with a knife, we finally put the bread in the oven, preheated to the highest setting, generally 250° (almost 500°F). We turn the oven off and let the bread bake in it for thirty minutes. Then we turn the bread out of its pan onto a cooling rack.

Let us taste it. There is a good chance that our bread will be heavy and dense, or not baked in the middle, too limp and full of holes. That is because each operation has its reasons, the importance of which bakers have learned to appreciate through experience. These reasons will make our future attempts crowning successes. And here they are.

To Work!

The first operation, the kneading, consists of combining into a dough the water, yeast, and flour, with a bit of salt to make the kneading easier and to improve the final taste of the bread. Why do we make a dough? Why is it elastic?

The main ingredient in bread, flour, most often comes from wheat, the only grain (or almost) that now allows for making leavened bread. Flour contains two principal components: starch granules, composed of two types of molecules, amylose and amylopectin; and various proteins, either soluble in water, such as the albumins and the globulins, or insoluble, such as the gliadins and the glutenins. We will see that these daunting preliminary chemical distinctions are significant enough, I hope, to make us overlook the heavy didacticism of the following explanation.

Let us begin by examining the proteins. If the dough produced during the kneading is elastic, that is because the insoluble proteins form the network that we call gluten. We must not think we are hopelessly far afield from bread. It is this network, stretched, that will form the thin partitions in the crumb. Air will be trapped when the dough is folded over on itself, and these walls will pull away from the gaseous cells.

Initially, the protein molecules are like chains folded back on themselves into a ball because of the intramolecular bonds already discussed: hydrogen bonds, between a hydrogen atom and an oxygen or nitrogen atom to which it is not chemically linked, or disulfide bridges, between two sulfur atoms.

Before kneading, these bonds are established between the atoms of a single protein molecule, thus producing its wound configuration. Kneading, however, separates the various proteins and gradually loosens the balls they form. As when spaghetti is poured from a saucepan into a colander or when waves sweep seaweed along the coast, the proteins become unwound by the movements of kneading, and they tend to form into lines.

When the proteins are aligned in this way, linked by hydrogen bonds, disulfide bridges, and perhaps other chemical bonds as well, the mass of dough becomes stiff, harder to work with, smoother, and more elastic. Many irregularities remain in the proteins, however. These intramolecular loops make up the springs that ensure the dough's elasticity.

So is the dough elastic or fluid? It all depends on the relationship between the concentration of glutenins and the concentration of gliadins. The glutenins are very large proteins that make the dough compact and fluid because they establish a tough, inextensible network; the gliadins, molecules about a thousand times smaller than the glutenins, ensure elasticity because they are more mobile and their loops reform more easily. Finally, the mechanical behavior of the dough also depends on the lipids present.

Why is wheat one of the only grains that make a good bread dough? Because its protein composition is such that the gluten formed is resistant enough to make a leavened bread. Wheat gluten is both elastic (it lets the bread expand) and viscous (it flows); hence it is described as "viscoelastic." It is because wheat contains more proteins good for bread making and less starch than other grains that the gluten is more durable when mixed with water.

Why Must the Flour Be Dry?

Enough about the gluten proteins. Let us move on to the starch granules that constitute the essential part of flour (70 to 80 percent). These spherical granules, two to forty micrometers in diameter, are, as we have seen, composed of two different molecules: amylose (20 percent) and amylopectin (80 percent).

Why do chemists call these molecules carbohydrates? Because their general chemical formula contains one carbon atom per each unit made up of one oxygen atom and two hydrogen atoms, as in water. But there are no isolated carbon and no water molecules in these molecules. Carbohydrates is thus a misnomer and should be dropped. Why do dieticians call these molecules glucides? Because the amylose and amylopectin of starch are both long chains, the links of which are the glucose molecule.[1] The difference between amylose and amylopectin is only the arrangement of the glucose groups in relation to one another. In amylose, the chain is perfectly linear, whereas in amylopectin,

---

1 Only before being linked, however. Strictly speaking, there are no glucose molecules left in amylose or amyylopectin, only glucose residues.

the chain branches. These two glucides are in the same family as cellulose, the structural compound of plants, formed by a chain of ten to fifteen thousand molecules of glucose. These are the building blocks of plant cells and bread.

And the bread builders? These are specialized proteins, present in small quantities but playing an important role. I am talking about enzymes. Enzymes are catalysts, that is to say, molecules capable of implementing chemical reactions without participating in them.

A simple example: when oxygen and hydrogen gas are brought together, they remain quietly mixed with each other. But if they get near a flame, they immediately explode, because the flame has catalyzed that reaction. Likewise, if a metal powder like platinum is introduced into a mixture of oxygen and hydrogen gas, an explosion takes place without the flame. The molecules of the two gases stick to the metal, split apart, and react. The metal serves only as a transient intermediary, and the reagents leave it in the state in which they initially found it. Another culinary example of catalysis: try to burn sugar with a flame. It will not burn, only caramelize. Now dip the sugar cube in ashes before trying to light it. This time, it will burn. Similarly, in the organic world, enzymes catalyze, promote, and accelerate biological reactions.

In the case of flour, it contains enzymes, the amylase group, that use water to detach long starch molecules from maltose, a molecule composed of two glucose groups and various other polysaccharides called dextrins, which serve as a nutritive substance for the yeasts. A remarkable twist of fate: flour contains precisely the enzymes that release the nutriments yeast needs from the dough, and yeast makes the dough rise.

Thus we can understand why flour must not be stored in a moist environment. The enzymes present in the flour would decompose it by using water from the atmosphere. Let us draw a lesson from this: for enzymes to act efficiently, let us hydrate especially the starch granules by kneading for a long time, and let us not skimp on the water. Thus, thanks to the enzymes, we will release the maltose that the yeasts will then consume, releasing the carbon dioxide that will make the bread rise.

The reasoning is straightforward. A long kneading produces a lot of maltose; a lot of maltose produces a lot of yeast growth; a lot of yeast growth releases a lot of carbon dioxide; a lot of carbon dioxide fills the alveoli in the bread with a lot of gas; and a lot of gas in the alveoli produces a loaf of bread that rises perfectly when baked.

Old Flour Makes Good Bread

Bakers know that flour that has been stored for a month or two makes better bread than fresh flour. Why?

We have just seen how kneading unwinds and aligns the proteins and how the proteins remain linked by hydrogen bonds and disulfide bridges that ensure the formation of intramolecular loops that serve as springs and give the dough its elasticity.

Disulfide bridges are bonds that are established just as easily between sulfur atoms in the same protein as between sulfur atoms in two neighboring proteins. Their reestablishment, after an extension, is compromised by the presence of thiol (SH: one sulfur atom [S] and one hydrogen atom [H]) groups on the neighboring proteins: the bond is established with the neighboring protein and not within the intramolecular groups; the loops do not reform after being stretched.

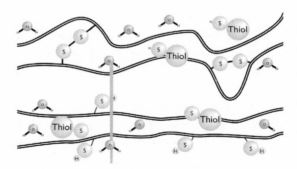

When a disulfide bridge is stretched with a thiol group in the proximity, there is a danger of a hydrogen atom passing over to one of the initially bonded proteins. The dough is more fluid than elastic. Aging the flour, which is accompanied by an oxidation of the thiol groups, gives dough better elasticity because the disulfide bridges reform better.

Let us note however that water can also give up hydrogen atoms to sulfur atoms when the dough is kneaded too much. But this danger is only a problem with mechanical kneading equipment. Bread makers usually get worn out kneading by hand well before this threat surfaces. And the actual practice of kneading? Begin by placing the dough on the far side of the kneading surface; unstick the dough from the surface and form a ball that you then bring toward yourself by folding it, trapping air in the process. Pound the folded dough firmly and repeat, flouring from time to time.

The Fermentation

Now that we know what hard work kneading is, let us take a rest and let the second step in preparing bread take place: the fermentation. This natural, spontaneous phenomenon is produced when the yeasts, mixed with the flour kneaded with water, can finally enjoy the pleasant environment that we have prepared for them.

There is a difference between wild fermentation, using sourdough (dough leavened with dough retained from the last round of baking) and fermentation using manufactured yeast, obtained by selecting yeast colonies cultivated on organic breeding grounds.

To keep things simple, we will first use baking yeast (not to be confused with baking powders, which do not have sufficient leavening power). During bread making, the yeast serves to lighten the dough by creating alveoli; it also gives it flavors and aromas.

Yeasts: the intrusion of living organisms into the dough. It is really a matter of single-cell microorganisms that proliferate when they have at their disposal specific compounds such as maltose or glucose. Beginning from these nutriments, the yeasts synthesize proteins and various other constituent molecules. Then they divide into two new identical cells, which divide in their turn, and so on. The higher the temperature rises (to the extent that the yeast is not destroyed), the more rapidly the yeasts develop. That is why they are first put in suspension in warm water; this awakens them.

During the first fermentation, which lasts about an hour, the yeasts ferment the maltose. The maltose is altered by the yeast enzyme, maltase, which breaks it down into two glucose molecules. These glucose molecules are then transformed into carbon dioxide, ethyl alcohol, various aldehydes, ketones, and other sapid and aromatic alcohols.

The fermentation of yeasts couples two reactions: the transformation of glucose $C_6H_{12}O_6$ into two molecules of carbon dioxide and two molecules of ethyl alcohol C_2H_5OH (Gay-Lussac reaction), coupled with the transformation that produces ATP, the molecule that serves as fuel for living cells.

Sourdough

We are not aware of them because they are so little, but yeasts—and various other kinds of microorganisms—are everywhere. Leave fruit out, and after a few days microscopic fungi, carried by the air, will naturally develop on them. Heat milk to a moderate temperature, and you will get yogurt, because the bacteria naturally present in milk will be nourished and will transform the milk into a gelatinous mass.

Similarly, you can prepare sourdough bread without adding baking yeast by simply using a mixture of flour and water that has been left to be colonized by the yeasts and bacteria that constitute bread's natural microflora.

Begin with some dough and add a bit of honey to it. After a day of fermentation in the open air in a spot that is moderately warm, mix this dough with the same amount of fresh flour and work in a bit of water and salt. Repeat this operation four times at about six-hour intervals. Then use this sourdough to make your bread. The wild *Saccharomyces minor* yeasts obtained will give it its slightly sour character.

White bread, the delicious white bread of baguettes and bâtards that we know in France, is a relatively recent development in human history. In the past, most bread was sourdough bread, and much attention was given to its preparation. In *L'art du meunier, du boulanger et du vermicelier*, published in 1771, Paul-Jacques Malouin (1701–77) writes,

> Sourdough bread is made from a piece of dough taken from one of the batches of the day; this is the principal sourdough, or the principal, and its weight varies from five to ten kilograms.

The baker adds flour and water to it in order to double or triple the weight. This sourdough, revived by kneading, is the first sourdough, which is left to ferment for six to seven hours. After another addition of flour and water, followed by another kneading, it becomes a second sourdough. Then once again the operations are repeated to form a completely achieved sourdough. After a fermentation of two hours, this serves as a culture for the dough. One obtains a light dough, bound and airy, by adding to the sourdough flour and water, mixing it, adding salt, cutting and folding it, and finally beating it.

Brewer's yeast only appeared in bread making in 1665, when a Parisian baker made the first yeasted bread. Until 1840 sourdough and cultivated yeasts were used in conjunction for the "French product." Then an Austrian baker introduced the use of yeast alone into France, and yeasted, or, as they are known in France, Viennese breads were born.

When Is Fermentation Complete?

The optimal temperature for fermentation for bread is 27°C (80°F). The yeasts would develop more rapidly at 35°C (95°F), but bitter metabolites would be released, and the dough produced would be stickier.

Through experience we know that the fermentation is complete when the volume has doubled and when we can poke a hole in the dough with a finger and it does not immediately close up again. At that point, the gluten has been stretched to the limit of its elasticity.

Why the Second Kneading?

When the fermentation is complete, we must go back to work. The function of the second kneading is to distribute the developed yeasts so that, during a second fermentation, there will be more of them available to release carbon dioxide.

The principle is the same as before, but after this second kneading, when the loaves are formed, we will slit the top of the dough with a knife a few times, so that the gluten network is not stretched to the limit of its elasticity just before it bakes. The bread will be able to expand under the pressure of the carbon dioxide without creating ugly tears in the crust.

The second fermentation, called the *apprêt*, or finishing, is the opportunity for the yeasts to use the sugars from the flour or those released by the starch and the amylase.

The Virtues of Carbon Dioxide

The carbon dioxide that yeasts provide is the same gas that is released from beer or champagne. It leaves the beverage in which it was dissolved as soon as the bottle is opened and the pressure diminished. In water, where it is dissolved in the form of carbonic acid, it stings the tongue, heightens the flavor, and acts as a mild bactericidal. Carbon dioxide is said to speed the transit of the bolus from the stomach to the intestines, which might explain why we can get intoxicated on champagne so quickly.

Finally, as we have seen, carbon dioxide is produced during the anaerobic fermentation (in the absence of air) of glucides by yeasts. Often this fermentation also produces ethyl alcohol, which gives bread its flavor, as well as various aldehydes, alcohols, ketones, and diacetyl, which contribute to the taste and aroma even before the baking, when Maillard reactions take place. But let us not get ahead of ourselves.

Why does carbon dioxide (and the air introduced during the kneading process) make the bread inflate? Because, like all gases, carbon dioxide expands when it is heated. In order to expand, it needs space, which it creates by pushing aside the dough, which is still soft before baking.

Thus it is important that the baking process not rigidify the gluten network too greatly before the gases present in the dough or formed during the baking can expand. Also the importance of long kneading becomes clear. The more the dough has been kneaded, the more finely distributed the gas will be, and the smaller the alveoli.

The Baking

The crucial moment has arrived, the moment of baking, which will crown our endeavors and give us the golden, sweet-smelling bread we have been waiting for.

During the baking process, the air introduced by the two kneadings and the carbon dioxide released by the yeasts expand. Simultaneously, the water and alcohol are vaporized and the yeasts' activity increases. At temperatures above 60°C (140°F), the yeasts cease all activity, and at temperatures above 90°C (194°F), the crust begins to form. Then, at 100°C (212°F), the water evaporates, and the vapor is distributed in the bread. The starch jells and passes from a semicrystalline state into an amorphous one, and the body of the bread forms. The gluten proteins are denatured by the heat; they coagulate after losing their hydration, and they form the rigid framework of the bread.

The water that evaporates leaves only via the surface of the bread, which dries and hardens; the crust is formed. In this regard, let us not forget to mention that the crust's color and aromas result from Maillard reactions. And let us note that, to obtain well-browned crusts, bakers toss a bit of water into their preheated ovens before putting in the bread. That is what they called the *coup de buée*, the puff of steam.

At what temperature should the bread be baked? We know that the baking temperature must not be too high, or the gases will have no chance to inflate the bread before the network of proteins rigidifies, or too low, or the water will remain in the bread despite the baking. Baking must be done between 220° to 250°C (428° to 482°F), some say at 230°C (446°F) for 15 to 20 minutes for baguettes. It is a question of balance. If the oven is too hot, and if there is no steam injection first, the crust forms before the bread inflates, but if the temperature is too low, the bread inflates before the crust forms, and the starch on the surface does not have time to form a network and the gluten doesn't coagulate; thus the bread falls again.

Why Does Bread Go Stale?

Going stale is not a matter of drying out. The concentration of water in the bread remains constant, but the starch molecules, which are irregularly distributed and bound to the water molecules, crystallize, expelling a portion of the water; the crumb becomes more rigid.

Why does well-baked bread rapidly become dry and stale? Why does stale bread become "fresh" again when heated in the oven? Why does the baker put fresh bread in the freezer to prevent it from going stale? Why does bread go stale less quickly in cloth or a closed box?

The explanation is clear if we remember that bread is obtained by baking a starch, that is, flour and water. If the bread is not baked enough, too much unused water remains. This water establishes additional bonds between the cellulose fibers; the bread hardens. If you heat it, you will break these hydrogen bonds, and the bread will become crispy again.

In the open air, bread goes stale by forming new hydrogen bonds. If it is not baked enough, putting it in the freezer prevents the excess water molecules from migrating and creating new bonds. Covering bread protects it from the air's humidity and prevents water molecules from penetrating it to create unwanted bonds. In well-baked bread, there are just the hydrogen bonds necessary to assure consistency and a good texture. This bread remains fresh longer, especially if it is enclosed in a bread box. Let us remember that!

Wine

The Mouthfuls Most Discussed Taste Best

Writing down "wine" is already to ask a question. Did I say *a* question? No, a thousand of them! Because of its complexity and diversity, wine escapes the closest analysis. We perceive the subtle odors, search for memories, often get lost there. Thus I will not proceed in my usual manner. With more modest goals than in other chapters, I will be content to try and describe this divine product, in order to appreciate it better. This is not just an intellectual exercise, because according to Grimod de la Reynière, "the mouthfuls most discussed taste best."

Because wine is a liquid, we can treat it differently from solid food when we taste it. Wine appeals first to the eye: we study its "robe." Then to the nose: we breathe in its bouquet. Finally to the mouth: tasting it many times, we confirm or alter our first impressions, we search for tastes and other odors, and we analyze their development and general harmony.

Tasting with the Eye

We scrutinize the wine's "robe" by gently tilting the glass and looking at it from above, so as to see a decreasing thickness of liquid.

The eye distinguishes many characteristics: the nuance (that is, the color); the highlights; the frankness (transparency, clarity, turbity); and the brightness (the luminosity, that is, the brilliant or dull character).

The eye provides much information to the one who knows how to use it. Is the color dark or light? Does it have nuances that remind you of other nuances detected at earlier tastings? Is the robe young and fresh or a little darkened with age? What does the colored disc indicate? Does the intensity of the robe extend to the edge of the glass, a sign of a quality product?

To describe these impressions, choose among the following terms for the robe: raspberry, intense, beautiful amber, yellow straw, light, limpid, brilliant; for the highlights: cherry, purplish, rosé, ruby, old rose, garnet, green, yellow-green; for the tears (more on these below): yellow, clear, viscous. . . . This list is not exhaustive, because wine is all poetry.

And this is also why our eyes can misguide us. Two scientist friends of mine, Gil Morrot and Frédéric Brochet, did some remarkable experiments in which they added red pigment to white wines and measured how professional wine tasters were misled. More recently, they observed how acidity was wrongly detected when green pigment was added to white wines.

The moral of this story? First, let us use all our senses to experience our food and not rely on just one. Second, let us be aware of the conditions influencing our food evaluations. And, finally, let us not forget poetry!

Drinking with the Nose

The nose, the sense organ for wine lovers! It perceives four features: the bouquet, the finesse, the aroma, and the development.

First of all, the bouquet can be more or less ample; "unassuming" or "powerful" are the usual qualifiers. Second, the finesse is an especially qualitative notion: wine can be common, even vulgar, or elegant, racy. The aroma corresponds to the wine's perfumes. A wine can be flowery because it has the odors of violets, for example, or peonies. It can be fruity, because it has the scent of raspberry, cherry, or plum. It can evoke the odors of wood, mushrooms (truffles, for example), or present animal smells. Development, finally, is essential. If a wine is too young, we will note that it is dumb or, on the contrary, aggressive; if it is too old, we will find it faded, stale.

To judge a wine's odors better, to track them, we must have an idea of what we can look for. The following list of descriptors should be useful: exuberant odors of red currant, raspberry, violet, flowers, ripe fruit, mushrooms, undergrowth, green wood, game, caramel, leather, smoke, tobacco, red fruits, black currant buds, roasted almonds, fresh fruit, green pepper; fruity, slightly acid, pleasant, wild, developed, present, rustic, complex, racy . . .

Search your memory for those terms that best correspond to the wine you are drinking and take advantage of the euphoria of wine tasting to overcome excessive modesty. Do not hesitate to be a bit individual: given the same wine, Western tasters will distinguish the aromas of dark and red fruits, whereas Japanese tasters will recognize the scents of seafood!

The Beginning of Ecstasy

In the end, it is the mouth that initially provides confirmation of the features identified beforehand by the nose and then allows us to proceed in identifying these four features: the strength in the mouth, which is relative to the degree of alcohol in the wine (light wine or fortified wine); the smoothness, which is the result of the glycerine, which makes the wine "fat," or results from the presence of sugar for mellow white wines (a red wine can be thin or fat; a white wine can be dry or sweet); the acidity, which makes a wine acerbic (halfway between incisive and piquant) when it is too pronounced or flat or soft when it is lacking; the body, which results from various biochemical factors, such as the degree of alcohol, the concentration of glycerin, the concentration of sugar and acidity, already mentioned, but which also takes into account other factors, like the concentration of tannins, which make the wine "keep" in the mouth, where its flavor lingers.

The body, a very synthetic impression, also corresponds to what certain wine lovers call the constitution. The constitution varies between full-bodied, powerfully built wines, and thin or hollow wines.

If such gustatory analysis allows for correcting the olfactory analysis that precedes it, tasting with the nose, after the first pass through the mouth, allows for perfecting this gustatory analysis: the scents rise from the throat to the nasal passages.

Finally, we must not forget to judge the astringency of the wine, especially at the tip of the tongue, its bitterness, its sweet or possibly salty (this is rare) character, and its finish: that is to say, after swallowing, the time during which the palate remains permeated with the various sensations experienced in the mouth. A good finish is the sign of an interesting wine. It is noted by measuring the number of seconds that the sensations persist. We say, for example, "this wine lasts five *caudalies*," if the sensations persist for five seconds. Certain exceptional wines create the impression of the aromas returning after a short time. Then we say that the wine "makes a peacock's tail."

Finally, Drinking the Wine

In practice, the taster can employ two different methods or employ them one after the other.

According to the first method, one drinks without swallowing a small quantity of wine. After placing the wine just behind the teeth, the tip of the tongue is immersed in it in order to determine the astringency (an impression of harshness, typical of tannic wines and which one can learn to recognize by chewing a rose petal), a possible sweetness (a sweet wine provides a clear sensation at the tip of the tongue, combined with an impression in the mouth more or less oily or thick), or acidity (impression of coolness on the sides of the tongue). Then, keeping the head back, one slightly parts the lips to breathe in a thin stream of air and aerate the wine. New "retro-olfaction" flavors appear; these are less volatile odorant molecules than those inhaled when projecting one's nose into the wine glass.

According to the second method, one chews the wine by turning it over in the mouth. In this way, the body and the fat of the wine are perceived.

A final note: when tasting several wines, one after another, it is standard practice to spit out each mouthful into a pail.

To describe the sensations in the mouth, we have the following list of adjectives from which to choose: winey, full, aggressive, harsh (synonymous with hard; the harshness of a wine is due to its bitterness or its tannin), pungent (more pejorative than rough), round, charming, well filled-out, voluminous, ample, supple, drinkable (wine that is easy to drink and refreshing), fleshy (wine that is so fortified and rich that it feels as though you could chew it; wines rich in tannins are often fleshy and are said to have a chew), frank, rich, lively, structured, fruity, flat, alcoholized, heavy, with a strong finish, with balanced aromas, solidly built (well structured in tannins), tannic in the back of the mouth, lingering, fat (for rich wines, full of substance, that fill the mouth well), with a good body, light (low in alcohol), smooth, silky (delighting the palate with a softness, similar to that of silk), heady, young, firm, unctuous, fortified (very rich in alcohol, going to the head), astringent (for wines with abundant tannins, not yet transformed enough, as we shall see, and lacking suppleness, like young Médocs), acid. For faults, we have the following: musty, maderized, oxidized, séché, flat, short (a poor finish, used for wines that provide too fleeting an aromatic sensation, not lingering in the mouth).

A wine taster's overall assessment is neither truly objective nor completely subjective. It is objective insofar as it cannot be positive if the wine has obvious faults, but it remains partly subjective because a wine may give an all-around good impression and still offer little pleasure.

Improve a Wine?

It is a great temptation, considering the price of wine, to buy cheaper wines and try to improve them.

Your wine is too light? A chemist would add ethylic alcohol to it. It lacks bouquet? Why not try a drop of cassis or vanilla extract? It lacks body? A little glycerin, for example. Not tannic enough? Add the necessary tannins. You like the taste of old wine? A little vanillin or even Madeira or port. You prefer a flowery wine? A hint of linalool. Drawn to the bordeaux? Try n-octanol and 2-methoxy-3-isobutylpyrazine. You prefer the leathery aromas of the burgundies? A little parethylphenol. And then there are still other compounds that are naturally present in wine but may be insufficiently evident in the bottle you have in your hand: ethyl methylanthranilate, paravinylphenol, wood lactones (compounds extracted by the wine from oak casks), cinnamon, nutmeg, acetaldehyde (the main aldehyde in all wines, found in large proportions in sherry, to which it contributes its characteristic taste). In light wines, a small quantity of acetaldehyde highlights the bouquet; in excess, it is undesirable, unstable, and causes oxidation.

I do not want to suggest that such manipulations are panaceas. First, only individuals who drink wine for pleasure can play such games, as it would be adulteration, that is, fraud, for wine merchants to do so. For trade purposes, wine is wine, not a mixture of chemicals. Second, some wines are works of art, gourmand Mona Lisas that chance attempts have little possibility of reproducing, and the bouquets of many products offered by true winegrowers are not reducible to simplistic formulas. The chemist must also consider his health and only use compounds uncontaminated by dangerous substances, and he must not abuse these trials; drunkenness waits to catch him unawares.

Your Own Fruit Wine

To understand why wine is such a complex crowning jewel, let us examine the preparation of an original wine made from raspberries, strawberries, black currants, potatoes. . . . One can, in fact, make wine beginning with almost any fruit at all.

Wine is produced through the fermentation of the sugar in grapes. One recipe consists of placing the grapes (or other fruit, or potatoes boiled with sugar and lemon juice, for example) in the presence of baking yeast in an open jar. As soon as fermentation has taken place, the liquid is poured into a container through a filter. Then it is tightly corked. Of course, the grape is the fruit that best lends itself to wine making.

Winegrowers carefully choose the yeast cultures they use. To make a burgundy or a Côtes du Rhône, they often use *Saccharomyces cerevisiae* yeasts, from the Montrachet culture, which ferment at between 18° and 25°C (64° and 77°F) in a few days. They make white wines at lower temperatures, because the yield in fruity esters is higher that way. To produce a sparkling wine, they resort to cultures that undergo a secondary fermentation, like *Saccharomyces bayanus*.

Similarly, the type and quantity of the sugar contained in the fruit's juice is fundamental. A must containing 20 percent sugar produces a wine containing 12 percent alcohol. From this perspective, the grape is ideal. Other fruits generally contain too little sugar, and it must be added before the fermentation.

Acidity, as well, plays a role in the development of the yeasts. The ideal fermentation takes place when the juice is acid (the pH between 3.2 and 3.6). At the end of fermentation, it is preferable to have a little acidity, because alkalinity dulls the wine. On the other hand, when the juice is too basic, there is the risk of undesirable substances being produced. That explains the importance of adding lemon juice at the beginning of fermentation in your attempts at fruit wines.

During fermentation, the higher the temperature, the more significant the extraction of tannins and color will be if the juice is in the presence of the pulp and stems. Nevertheless, at too high a temperature, the yeasts produce substances (decanoic and octanoic acids, and corresponding esters) that neutralize their properties, feed on them, and kill them off. The fullest, darkest, most tannic red wines, which have the longest possible lives, remain in contact with the skins for ten to thirty days (the skins contain both the pigments called anthocyanins and the tannins). On the other hand, the lighter red wines are separated from the skins at the end of just a few days. White wines are obtained by fermenting the juice alone.

When the fermentation has ended, that is, when bubbles have ceased to form (strictly speaking, fermentation takes place in a closed but not air-tight vessel, in a cool, well-ventilated place), siphon off the juice and place it in a sterilized container. Then add a small quantity of a solution of 10 percent sodium metabisulfite, which will protect against oxidation and clarify the product. In the winegrowing industry, sodium metabisulfite is replaced by sulfurous anhydride or ascorbic acid, which are aseptic. Some winegrowers also perform a "flash pasteurization" at this point. In order to kill the yeasts, they bring the wine to 80°C (176°F) for about thirty seconds. The yeasts are inactive above 36°C (96°F), and the enzymes are destroyed at temperatures above 65°C (149°F).

Now, "*collez*" (glue) your wine in order to clarify it. After fermentation, wine contains matter in suspension, which can cause cloudiness in the bottle. The *collage* (gluing) clarifies the wine and eliminates some of its undesirable characteristics. The most common clarifiers are egg white, gelatin, bentonite, fish glue, and casein. Egg white, with its positive charge, eliminates negatively charged matter (for example, undesirable tannins or anthocyanins), whereas bentonite, negatively charged, eliminates positively charged matter (proteins and other organic materials).

Finally, sometimes wine is put in wooden casks (generally oak) to age, where the alcohol extracts tannins from the wood and gradually react with them, producing various aldehydes and then odorant molecules like vanillin (which is why I encourage you to put vanillin in certain slightly weak wines; also see the chapter on alcohols).

Why Does Red Wine Darken as It Ages?

Red wine owes its color to vegetable pigments called anthocyanins. When red wine ages, the anthocyanins react in a variety of ways. Some react with other colorless but bitter compounds, which are also present and collectively designated by the name of tannins. This reaction does away with the bitterness and astringency of the tannins, which precipitate (thus forming the deposit in certain old wines), and improves the taste of the wine.[1] A wine that is too green and astringent becomes smoother and more supple. As the wine continues to age, this reaction makes the red of the anthocyanins disappear and lets the dark tannins become visible.

Why Does White Wine Lose Its Green Highlights as It Ages?

The color of white wines is especially due to the presence of quercitin, a molecule that becomes brown as it is gradually oxidized. Initially, young white wines have a green tint because of the chlorophylls that are extracted during fermentation, but gradually the quercitin dominates the robe and enriches it. Chlorophylls also generate odorant molecules as the wine ages.

How Should We Keep Wine?

The answer to this question is based only on experience; science, alas, has not studied it. Common wisdom says that the temperature of the wine cellar should remain constant, between 8° and 15°C (46° and 59°F) all year round. Wine bottles should be kept in the dark, resting on their sides, and should not be moved, and the air of the cellar must be good.

---

1 Many similar processes have been discovered in recent years, in particular by two marvelous chemists, Véronique Cheynier, at the INRA Center in Montpellier, and Raymond Brouillard, at Strasbourg University.

All these recommendations warrant verification, because they contradict other stories, such as the one about those bordeaux wines that were improved, it is said, by a round-trip to the Indies. So much for rest and cool air!

The fact remains that the ultraviolet rays of sunlight should be avoided, because they stimulate chemical reactions that alter the wine. Brown wine bottles offer better protection than green ones. Some of the photochemical effects from ultraviolet light may be reversed, it is claimed, by keeping the wine in total darkness for a few months.

Why Does Wine Weep?

The tears of wine, which make wine lovers say with a knowing air, "Ah, this wine has legs,"[2] are due to the presence of alcohol in the water, but their movement also depends on the presence of glycerol (the chemical name for glycerin). The tears are more evident to the degree that the alcohol level is raised. Often the legs that descend when the glass is inclined are confused with the real tears, which form spontaneously at the ambient temperature when the glass remains immobile.

In 1855 the British physician James Thomson proposed that the tears of wine resulted from the differences in wettability between the wine in the glass and the wine that was lower in alcohol content because of the alcohol evaporating.

The phenomenon works as follows: in a still glass, the liquid rises spontaneously along the walls, forms a thin crown a few millimeters above the wine's surface, and then descends again in the form of tears that remix with the wine. Since the crown is replenished by a rising tide of liquid, the tears remain for several minutes after wine has been poured into the glass.

If the glass only contained pure water, in a moist atmosphere, these tears would not exist. With wine, the crown is regularly replenished. The water-alcohol mix wets the glass, forming a meniscus at the extremities of which the

---

2   And, in saying so, they are wrong, for legs are different: tears are when drops form spontaneously on the glass, and legs are when the same pattern is formed only when you tilt the glass, before setting it upright.

alcohol evaporates more rapidly than the water. The impoverishment at the extremities prompts an aspiration of the wine, from the bottom of the glass to the top of the meniscus. Simultaneously, the solution weaker in alcohol redescends in the glass.

When the tears redescend, the nearly pure water that they contain comes into contact with the wine that remains in the glass, and sometimes the difference in wettability makes it so these two liquids do not mix. It is the same effect as when a sink is drained after doing the dishes. The detergent sometimes remains clinging to the enamel, so that if drops of water are applied to it, they do not wet the sink. Here it is the alcohol that plays the role of the detergent (or the surfactant, to borrow the term used with regard to emulsions: mayonnaise, béarnaise sauce, etc.) with its CH_3CH_2 group, insoluble in water, on one side, and its hydrophilic group OH on the other.

Currently, the physical chemists who have studied the tears in wine have concluded that the alcohol and water in wine do not entirely explain the phenomenon. Glycerol, especially, noticeably alters their dynamic. We have already seen that the sweetness and slippery texture (the viscosity) of glycerol make it an interesting compound in wine. What we have not yet seen is that this compound is produced by the noble rot (the fungus *Botrytis cinerea*) that, under certain conditions, attacks grapes, damaging their skins and thus causing the water they contain to evaporate. Enriched in glycerol, such grapes produce sweet, smooth wines.

Is It Necessary to Let Wine Breathe Before Drinking It?

The authors of wine books are divided on this important question, and, once again, science offers little help in resolving it. The best rule seems to be to open the bottle a bit in advance and taste it. If the wine is a little rough, then aerate it by decanting it into a carafe. If not, leave it in its bottle to keep oxidation from deteriorating it.

This method has the advantage of revealing if the ideal temperature for consuming the wine has been attained (red wines with strong aromas are served warm, so that the volatile aromas are more easily released; nevertheless, excessive heat should be avoided, as the alcohol will volatilize in the air, and the wine will become sweet). Be careful in handling light wines; they deteriorate more easily than aged, heavy wines.

And now let us raise our glasses with Rabelais, and, "Drink!"

The Alcohols

How Can We Distill Alcohol?

In the past, distillers set up shop on the outskirts of villages with their carts and their copper stills to distill cider, wine, and the fermented juice of various fruits: pears, apples, plums. The principle behind distillation is simple. Since ethylic alcohol boils at 78°C (172°F) and water boils at 100°C (212°F), alcohol is separated from water by heating a mixture of the two substances; the alcohol, which evaporates first, is condensed in a coil, while the water remains in the vat.

In practice, the operation is a bit more complex, because the aim is to recuperate not pure alcohol but flavored alcohol. In addition, the methanol, or methylic alcohol, must be eliminated by eliminating the first distilled fractions; this alcohol is toxic and, most important, causes blindness (nevertheless, it contributes to the bouquet when it is present in weak concentrations in certain white alcohols).

It is especially important to know that a high-quality brandy can only be obtained beginning with white wines that are quite acid, have a very light bouquet, and are low in alcohol content, because distilling strongly flavored wines produces too heavy a brandy.

In addition, it is important that the distilling vat be copper. Copper atoms fix the fatty acids in the wine and also capture the sulfur of the sulfur dioxide often present in white wines.

If distillation were not forbidden, anyone could easily practice it at home. All you have to do is put the mixture to be distilled in a pressure cooker, connect a length of pipe over the safety valve, and make cold water run over the pipe to condense the distilled vapors. One or two runs in succession, eliminating the first and second fractions in which various toxic products are concentrated, will procure an alcohol of the desired degree.

Improved Whisky?

Once the alcohol is made, its taste can then be improved by letting it age in bottles in which sticks of dry wood have been placed (in eastern France, hazel wood is often used). (Better still, the wood in question can be heated briefly over a fire before being placed in the bottles. This operation, also carried out by barrel makers who heat their staves, causes other interesting compounds to appear.) The acids in the brandy gradually break down the lignin in the wood into phenol aldehydes, which are then oxidized into phenol acids. The brandy becomes less acidic while at the same time aromatic compounds, such as synapic, syringic, vanilic, and ferulic acids, appear.

Why dry wood and not green wood? Because green wood contains aesculin (bitter), which is gradually transformed into aescutin (sweeter) when the wood dries.

Since compounds like vanilla are present in aged alcohols in contact with wood, why not speed up the aging process by adding these compounds directly to young alcohols? Adding a few drops of vanilla extract to whisky, for example, will make it more full-bodied—but stop before the whisky smells like vanilla. Similarly, you can add a very small amount of cinnamon, since cinnamic aldehyde is formed in the same process as vanillin as alcohols age.

Cold Distillation

Another distillation method, less well known but perhaps even simpler than the one I've described, consists of placing the mixture to be distilled in a freezer. When it freezes, the water forms into a block of ice, separating itself from the alcohol and the other compounds that remain in the liquid phase.

Alas, it is also against the law to proceed in this fashion . . .

Why Does Alcohol Make You Drunk?

The compound commonly referred to as alcohol, which chemists call ethylic alcohol, or ethanol, is only one member of a huge chemical class of alcohols. In its pure form, it is a colorless, odorless compound that burns the tongue.

From its chemical formula, CH_3CH_2OH, we can locate its alcohol function in the OH group, which replaces a hydrogen atom in ethane (a compound with the formula CH_3CH_3).

Why the name "alcohol"? Because the Arab word, *al Kohl*, means "fine powder." Actually, the Egyptians tinted their eyelids with an inorganic compound, sulfur of antimony, which they ground in order to apply it. Then, the name was given to the essence of anything at all, notably liquids obtained by distilling wines, when this operation was invented by Avicenna in the tenth century.

Why does alcohol make you drunk? Because it stimulates the brain, which frees the cortex of inhibitory controls; that explains the excitement of drinkers, at least in the first stages of what health workers call "alcoholic intoxication." More precisely, alcohol works by interacting during neurotransmission. The brain cells called neurons function by receiving information from other neurons, by calculating the sum of activations and inhibitions, and by stimulating neurons further along in the system according to that calculated sum. A neuron activates other neurons by releasing neuromediating molecules that attach themselves to the receptor molecules of neurons further along.

The neuromediator with which alcohol interacts is gamma-aminobutyric acid, or GABA, which acts as an inhibitor. By attaching itself to its receptors, GABA deforms them and facilitates the entry of chloride ions into the neuron, which becomes less excitable.

On the other hand, when it attaches itself to the GABA receptors, alcohol facilitates the fixation of the neuromediator, so that the neurons further along in the system are less inhibited.

Now knowing the dangers that lurk for us in alcoholic beverages, let us remain temperate. . . .

Jams

Why Does Lemon Juice Make Jams Set?

Jam? Preparing it is so simple that we could leave it to children if they did not run the risk of getting burned: heat a mixture of sugar, a trace of water, fruit, and seal it in a canning jar. And there you have it!

You may encounter a few difficulties in the details, however, not from the point of view of conservation but regarding consistency. How to obtain jam that holds together? Why do some fruits make better jam than others?

The key to jam is a long molecule called pectin, present in the walls of vegetable cells in varying proportions. This is the jelling molecule. Composed of a chain of groups of hexagonal rings with five carbon atoms and one oxygen atom bound by short segments, pectins, like proteins, are kinds of long threads that bear COOH acid groups capable of ionizing, that is, the hydrogen atom can lose an electron.

This ionization is important for making jam because, when it takes place, the pectin molecules all have the same electrical charge and repel one another.

To form the gel that jam becomes through linking the pectin molecules, this repulsion must be avoided. The pectin molecules, separated from the fruit by heating it, must be allowed to reassociate into a three-dimensional network that fills the whole container.

Thus we can understand what conditions jam needs to jell successfully. The fruits must provide a sufficient quantity of pectin, the sugar content must be high, so that pectin association is promoted (see later), and the environment must be acid enough for the acid groups in the pectin not to disassociate and for the electrostatic repulsion between molecules to be kept to a minimum.[1]

Let us draw some conclusions from this analysis. First of all, the mixture of sugar and fruit must cook enough for the pectin to be extracted from the cell walls. The sugar, which must be well heated, pulls the water out of the cells into the surrounding syrup (through osmosis). It thus damages the cells, which further releases the pectin molecules. Since the sugar increases the boiling temperature for the mixture (pure water boils at the temperature of 100°C [212°F] but a mixture of one liter [33.8 ounces] of water and 900 grams [31.75 ounces] of sugar does not boil until it reaches 130°C [266°F]), it promotes the extraction of pectins as well.

The quantity of sugar must be significant because, even in an acid solution, pectins do not jell easily; they bind to water rather than among themselves. If sugar is added, it attracts the water molecules and leaves the pectin molecules unattached. Thus they participate in a group marriage, and gel appears. Some fruits do not contain enough pectin to form a good gel on their own (blackberries, apricots, peaches, strawberries) and must be supplemented with fruits in which pectin is abundant (grapes, apples, most berries). Finally, fruits that are not naturally acid must be supplemented with lemon juice, which deters the ionization of the acid groups of molecules in the pectin and thus their repulsion.

How Much Pectin?

Jam lovers are well aware of the fact: jams that are too firm are rarely good. Why add pectin to jam? Although it helps to preserve it, can it nevertheless do harm? In exploring the relationships between consistency and taste in jams, physical chemists at the INRA taste research laboratories in Dijon determined a few methodological ingredients for a good strawberry jam. The results can easily be applied to other fruits.

Traditionally, as we have seen, strawberry jam is made by heating the fruit in a mixture of sugar and water. After boiling it for a few minutes to let the excess

---

1 More precisely, the medium's pH must be around 3.3.

water evaporate and to kill the microorganisms that are present, the preparation is poured into sterile jars. Should the pot be covered? Should the mixture be heated slowly or at a rapid boil? Will poor-quality strawberries make a good jam nevertheless? What is the actual effect of adding a jelling agent? These important questions motivated the studies of the physical chemists in Dijon. If the consistency of commercial jams seems right, commercial products are well-known—and perhaps inaccurately—for lacking the flavorful characteristics of our grandmothers' jams. Where might the commercial jam industry's error lie?

Knowing that certain products called hydrocolloids, used to increase the viscosity of foods, reduce their taste and odor, the physical chemists in Dijon studied first the relationship between the gel in jams and the odorant compounds present. Many types of pectin are used by the farm-produce industry. Generally, highly methoxylated pectins with many lateral groups—called methoxyl—serve as gels for foods high in sugar, and pectins low in methoxyl are used instead in foods low in sugar. The INRA researchers thus compared five samples of jam containing very methoxylated pectin, at different concentrations; five samples with unmethoxylated pectin, at various concentrations; and a control sample in which the pectin came only from the strawberries.

Done under standardized conditions, the jam evaluation consisted of two parts: a chemical analysis of the volatile compounds and a sensory analysis, during which selected tasters described the products given to them with the help of twenty-five terms, defined preliminarily, including ten attributes for aroma and three attributes for taste. For each sample, the tasters also noted their assessment of the jam's consistency in the mouth. The tasting took place in a room lit with red lights, so that the color of the different samples (varying accord to the type of preparation) could not influence the gustatory assessment. The tasters were only given unlabeled samples, and each jam was presented twice and in random order.

The preliminary chemical analyses, in which thirty-one volatile compounds capable of contributing to the flavor were identified, showed that the concentration of these products differed greatly from jam to jam according to the fruit lots. The quality of the jam depended heavily on the quality of the fruit used in it.

In addition, during a preliminary evaluation of the jam's consistency by the tasters, it was verified that the tasters' responses were consistent, and two unexpected phenomena became apparent. All the tasters preferred the jams that did not contain highly methoxylated pectins, and the ideal concentration was in the neighborhood of the concentration generally used by the jam-making industry.

The next step involved determining the relationship between sensory perception and the presence of pectin. Thus it was discovered that increasing the concentration of highly methoxylated pectins enhanced the consistency and viscosity but diminished the notes of sweetness, acid, and caramel. The chemical analyses showed that only seven volatile compounds analyzed had notably diminished in concentration (the compound called mesifurane, which contributes a note of caramel, and various flowery or fruity esters).

With the pectins low in methoxyl, on the other hand, the oral consistency also improved with their concentration, but three times more pectin had to be used than with the high methoxyl pectins to obtain the same consistency. The jury did not note sensory variations in comparison with the control jam, even though chemical analysis registered an increase in the concentration of many fruity esters.

What to conclude from these studies? That the addition of pectins makes jam more firm but reduces its gustatory qualities. How? We know that a substance is only sapid or odorant if it circulates very well around the taste buds or the receptors in the nose. If aromatic compounds are bound to pectin molecules, thus blocking that circulation, the olfactory qualities are reduced. This explanation has been corroborated by experiments in which the volatile compounds were extracted by stirring the jam over the course of its preparation. Chemical analysis detected, in the vapors, many more compounds and in much higher quantities that when the jam was slowly simmered, which confirms that the bonds between the pectin and the volatile compounds are weak.

Finally, since the gustatory quality of jam depends greatly on the presence of weakly bound volatile compounds, the researchers wanted to study the influence of preparation conditions on the products' qualities. In the end, they observed a considerable loss of aromatic compounds through evaporation.

In other words, you will make good strawberry jam if you take the following advice: (1) choose good-quality strawberries; (2) add pectin only when the pectin in the fruit is insufficient (with cherries, for example); (3) do not stir the preparation too much while it is cooking; (4) heat gently in order to extract the natural pectins from the fruit and to avoid eliminating volatile compounds through either too strenuous stirring or extraction by the water vapor; (5) if possible, recuperate the vapors, and condense them, eliminating the water and returning the treated condensates, rich in aromatic compounds, to the jam before pouring it into jars.

Tea

How Long Should Tea Steep?

In Southeast Asia, tea leaves were chewed or infused in prehistoric times. Tea has been cultivated in China since the fourth century before our era, and its use was transported to Japan in about the sixth century. But if the practice of infusion is universal, all herbs and plants are not endowed with the same capacity for releasing scents and flavors.

Orientalism and, it must be confessed, a certain perfectionism about tea and its preparation, established its use in our countries, where we have not entirely forgotten that country people have made infusions from plants since time immemorial: mint, linden . . .

Let us give in to the tea craze. How long should it steep? Some tea drinkers recommend letting it steep longer than is necessary for extracting all the color, because certain flavors are released from the vegetable substances more slowly than the colorants. No doubt that is true, but only up to a certain limit, in particular the one that corresponds to the extraction of the tannins, substances that are bitter and astringent.

If this limit is passed, the solution of putting milk in the tea remains, but . . .

Tea in Milk or Milk in Tea?

When preparing tea with milk, should you pour the tea into the milk or the milk into the tea? Naturally, this is only a problem for those who, like the English, mix tea and milk together, but its answer may explain why our friends across the Channel are such tea enthusiasts. Prepared according to their method, tea loses its natural bitterness.

Even without having a taste for tea, we must recognize its great delicacy. Its light bitterness allows its delicate flavor and subtle scents to come through. How to retain those latter characteristics without the former? That is the milk's role, no doubt added originally for its natural sweetness, then embraced for the antibitter properties it possesses.

Tea is bitter because it contains tannins, those same compounds that give certain wines their astringency or even a marked bitterness, those same molecules that make a rose petal seem bitter if you put it in your mouth. Milk, on the other hand, contains many proteins, long chains folded back on themselves, that sequester the tannins. They bind themselves to them, destroying the bitterness.

An easy test of this is to add cold, raw milk to cold tea infused for a long time: the bitterness disappears. This same experiment, however, fails "hot," because the heat denatures the proteins, that is, it unravels them and deprives them of their sequestering properties. If tea that has steeped too long is added to milk that has boiled, the bitterness remains. Still worse, the taste of cooked milk masks the tea's flavors!

We now have all the elements we need to answer the initial question. If you add milk to very hot tea, its proteins will be denatured, and the tea's bitterness will remain. On the other hand, hot tea added to cold milk will lose its bitterness because the final temperature of the mixture will not be higher, at least at first, than the temperature at which proteins are denatured, and the proteins will sequester the tannins.

Change Tea's Color?

As long as we are adulterating tea, let us mention lemon. Why does its juice make tea lighter in color?

Does it, too, contain proteins that sequester the tea's colorant molecules? No, the explanation is of another order, more chemical than physical. Let us notice, first of all, that the tea to which lemon juice is added does not become colorless, or even yellow, like the lemon juice. Its red color turns to orange, because its red pigments are weak acids (acids are molecules that contain a hydrogen atom capable of dissociating under certain conditions). In the presence of lemon juice, that is to say, a stronger acid, the yellow color of the nondissociated form becomes apparent.

By adding bicarbonate to tea—and I do not guarantee the gastronomic results of this experiment—we obtain the reverse effect. An intense brown color develops as a result of the dissociation of the acid groups and the appearance of the other dissociated form of the pigments.

How Can We Not Spill the Tea When Pouring It?

The "teapot effect" is one of the most disagreeable phenomena encountered in cooking. With certain teapots, the one pouring knows in advance that the boiling liquid will spill on the knees of the guests or at least on the carefully washed and ironed tablecloth.

Physicists who recognize this effect have found an answer, but it is a Pyrrhic victory: they try teapots out before buying them. The effect was studied by Marcus Reiner in 1956 at the Technical Institute of Israel. Then, in 1957, Joseph Keller of New York University explained the phenomenon.

In physics, the flow of a liquid is characterized by the current lines, which are tangential to the velocity vector of water. More concretely, you can form an image of these lines by putting small colorant particles into a flowing liquid; the streaks of color are the current lines.

When water flows over a horizontal surface, the current lines are horizontal and parallel, but when the liquid encounters an obstacle, the lines draw together and the speed of the liquid increases; simultaneously, the pressure diminishes. This increase in speed is well known to all sailors. When a current rounds a point, the water accelerates ahead of the point.

The decrease in pressure, imperceptible to the sailor, becomes evident, alas, when you pour tea. In passing near the lower edge of the spout, the current is pulled downward by the weight of the liquid, so that it accelerates and its pressure decreases.

The pressure decreases at the edge of the spout, did we say? Since liquids have a tendency to be displaced from zones of high pressure to zones of low pressure, the tea that accelerates is plastered to the side of the teapot. That is what scientists call the Bernouilli effect, and it lets them make a liquid flow the length of a long glass tube. In the case of tea, the liquid faithfully follows the contour of the teapot . . . and ends up on the table!

Cold and Cool

Cool, but How Cool?

How to keep fruits and vegetables for a long time? By putting them in a cool place as quickly as possible, by isolating the ones that are already damaged, and by carefully cleaning the containers where they are kept. Of all the benefits of science and technology, one of the most misunderstood—because it is so familiar—is refrigeration.[1] Nevertheless, only the carefully considered use of refrigeration produces good results. Here is a way to use it that owes much to the findings of the team of agronomists at the INRA research center in Monfavet and to the work entitled *On Food and Cooking*, by Harold McGee, mentioned earlier. This book is a survey of everything related to food and its culinary transformations.

What we hope to avoid, through cold temperatures, is, for example, the degradation of plant tissues. Ideally, fruits and vegetables should be consumed straight from the garden, but as long as cities are not transformed into immense fields, we face the need to conserve our foods.

The composition of foodstuffs changes considerably within a few hours of being harvested because the plant cells continue to function even though they are no longer receiving water from the plant's roots. Corn and peas, for example, lose up to 40 percent of their sugars in six hours at room temperature. What is worse, asparagus and broccoli, once picked, use these sugars to synthesize indigestible woody fibers. It is not a silly sentimental illusion to believe that the taste of fresh vegetables is very different from that of vegetables left sitting in the pantry.

---

1 It seems so natural to us today that we even forget how conserving food by canning it, or *appertisation*, named for Nicolas Appert, constituted a health revolution.

Cold preserves this taste of freshness, slowing decomposition and preventing degradation by microorganisms. Why? First, because plant cells live more slowly in the cold than at the ambient temperature, biochemical reactions take place more slowly. Microorganisms are slowed down as well, so they proliferate less and degrade vegetables to a lesser degree.

On the other hand, fruits and vegetables cannot stand the cold at all. Certain tropical vegetables in particular are especially sensitive to it. Bananas, for example, are damaged by their enzymes, which brown the banana skin. Avocadoes darken and do not ripen at temperatures below 7°C (44°F). Lemons and other citrus fruits get spots. Pineapples, melons, tomatoes, cucumbers, and green peppers keep better at 10°C (50°F) than at lower temperatures. Potatoes get soft at temperatures below 4°C (39°F) because the starch in them continues to turn into sugar. Most other vegetables—carrots, cabbage, greens, and so on—keep well at about 0°C (32°F). Their cells contain salts that prevent freezing according to the same phenomenon that lowers to –17°C (1°F) the temperature of a mixture of ice and salt.

The Big Chill

Freezing completely halts respiratory reactions in vegetables, but it kills plant tissue. Water in the cells forms ice crystals that pierce the plant cell walls and membranes. During the freezing process, vegetables wilt because the broken walls and membranes no longer keep the cellular mass rigid. To avoid this inconvenience, the cooling must take place as quickly as possible. In that way, the ice crystals that appear remain small and numerous.

One more precaution: freezing considerably reduces enzyme and chemical activities, but it does not block them completely. The only way to terminate all activity is to blanche the food product. Quick immersion in boiling water inactivates the enzymes; subsequent immersion in cold water stops the cooking process and weakens the cell walls.

Fruits, however, are rarely blanched, because blanching makes them lose their flavors and textures. The enzyme action that turns fruit brown is better prevented with the help of a sugar, alcohol, or an ascorbic acid solution.

Vinegar

The Acid of Alcohol

Since Louis Pasteur's time, we have known that vinegar is formed through the fermentation of ethylic alcohol by a single-cell fungus related to yeast, *Mycoderma aceti*. In conditions of limited acidity, with certain concentrations of alcohol, and in the presence of nutritive compounds such as the proteins present in wine, these mycoderms develop and form a grayish veil, sometime as fine as silk, sometime more solid.

The fungus absorbs oxygen from the air and fixes it on the alcohol, transforming the alcohol into acetic acid, which, as a solution in water, constitutes vinegar.

Mycoderms like acid products and develop better if the environment is initially a bit acid. To make vinegar, adding a bit of already formed vinegar to the wine to be transformed is recommended. This addition has the advantage of preventing the wine from being colonized by the "flower of wine," another microorganism that prompts wine to spoil.

Must You Have a Mother to Make Vinegar?

Often recommended as an aid in making vinegar, mother of vinegar is composed of acetic mycoderms that have not penetrated the mass of vinegar . . . and that therefore act in a harmful fashion. Instead of transforming wine into vinegar, they destroy it, consuming the oxygen in it, because, in solution, it lacks them.

What is worse, mother of vinegar destroys the odorant compounds that give vinegar its odor. The conclusion is incontestable. When making vinegar, avoiding mother of vinegar is absolutely necessary. Only the surface veil is beneficial.

Vinegar is produced in various ways, but the Orléanais method is done with the help of a row of casks. Wine is regularly added to the highest cask, and vinegar simultaneously decanted from the lowest cask. More precisely, to make 230 liters (around 243 quarts) of vinegar, 8 to 10 liters (around 8.5 to 10.5 quarts) are decanted each week, replaced by an equal amount of wine. The operation must take place in a half-full container, so that there is maximal exposure to air. The temperature must be consistent, and the veil (which forms spontaneously) must not be damaged by the addition of wine.

Vinegar can be made with various fruits, raisins, honey diluted with water, cider, fermented pear juice, berries. . . . But the best vinegar is made with good wine. And, as everyone knows, aromatic vinegars can also be made with various herbs, such as tarragon.

Let us also note that balsamic vinegar, made in the Modena region of Italy, is the only vinegar that harmonizes with the wine served during a meal. This vinegar is produced beginning with white grapes and sugar. After harvesting, and with the first signs of fermentation, the must is drawn from the vats, filtered, and slowly boiled. Then it is filtered once again and passed from smaller to smaller casks while the acetification takes place and the liquid becomes concentrated. Try it with walnut oil to accompany a salad garnished with sliced truffles! It is more expensive than standard vinegar, but what a pleasure not to ruin the taste of the wine when you eat your salad.

Kitchen Utensils

How Can We Rejuvenate Silverware?

Silver place settings, the treasures of our grandmothers, ornaments to our tables, a pleasure to the eye, have a serious inconvenience. They tarnish. If they come in contact with egg? Their radiance seems irremediably lost. If they are washed in a sink that contains less noble metals? They darken as if they were too fine to endure contact with commoners.

How to recover them? The solution is simple. Remedies abound, but some are not reliable. I offer you here my full assurance of the perfect effectiveness of the following two remedies.

The first possibility involves the use of hydrogen peroxide. Actually, silver turns dark only because it is oxidized, generally by sulfur. Abundant in eggs, sulfur binds to silver in an insoluble compound of silver sulfide. Hydrogen peroxide continues the oxidation, transforming the insoluble sulfur into soluble silver sulfate. Consequently, this remedy should be reserved for solid silver place settings only.

A second possibility, hardly more complex, is effective for silver-plated settings. The sulfur can be dissociated by electrolysis, and the silver plating can thus be preserved.

To do this, cover the bottom of a plastic container with aluminum foil. Add hot water and a tablespoon of cooking salt. Then place the tarnished objects in the container is such a way that they are in contact with the aluminum. Thanks to an electrical circuit composed of the conducting solution (salt serves to make the water a conductor), the aluminum, and the silver, the following chemical reaction takes place. The aluminum loses electrons, which flow to the metallic silver. On the surface of the place settings, the silver bound with sulfur captures these electrons, recovering its metal form, while the sulfur is released into the solution, migrates toward the aluminum, and forms aluminum sulfide.

You can speed up the process by using water that is almost scalding.

Why Beat Egg Whites in Copper Bowls?

Is it necessary to use copper bowls when beating egg whites into stiff peaks? Copper is said to make egg whites stiffer than other materials do, and furthermore one of the plates in Alembert and Diderot's *Encyclopedia* depicts, in an illustration of a kitchen, a copper bowl for beating egg whites.

Scientifically, the jury is still out. Apparently, egg whites beaten in a copper bowl are stiffer than those beaten in other containers, but what is the reason for that?

The *"cul-de-poule,"* that hemispheric copper bowl reserved for beating egg whites in large kitchens, has the advantage of never coming in contact with fat, which, as we have seen, inhibits the bonds between the egg white proteins. It could be the perfect cleanliness of these bowls that produces those perfectly beaten egg whites.

The significance of the copper has been tested since the beginning of the century, when it was observed that one egg white protein, conalbumin, bonded to metallic ions and thus became much more resistant to denaturation. It was then assumed that if the conalbumin bonded to the copper in the bowl where it was beaten, overbeating would become more difficult. It has been verified that conalbumin does bond to copper (the color of egg whites beaten in copper is different from that of whites beaten in iron, for example, because the conalbumin-copper and the conalbumin-iron complexes have different light absorption properties), but many studies must still be done to confirm that the sequestering of metals is really responsible for stabilizing egg whites. And even with this hypothesis, it has not been established that egg whites beaten in copper withstand cooking better; nothing replaces experimentation.

Nevertheless, let us mention that copper must be handled with caution. It is so toxic that water placed in copper and left in the open air is not colonized by ambient microorganisms. Moreover, what matters most of all is the perfect cleanliness of the container. The presence of fat, as we have seen with regard to soufflés, interferes in forming stiff peaks. If you have any doubt and suspect that fat is present, clean the bowl by rubbing it with salt and vinegar or with a lemon quarter.

Why Cook in Copper Saucepans?

Copper saucepans seem like a luxury. Are they really? That is not clear, because copper conducts heat very well. In any one spot in the pan, all the excess heat is rapidly dissipated because the heat spreads quickly into the rest of the utensil. Thus a copper pot reacts very rapidly to variations in temperature, which ensures cooking by the entire surface of the pan, bottom and sides, without "hot spots," points that overheat and trap molecules, carbonize them, and give the dish a burnt taste. With copper, it seems that the temperature is better controlled, that it can be adjusted at will, without too much lag. This trait is indispensable for the most delicate sauces and for slowly simmered dishes.

To avoid the toxic contact of verdigris, copper utensils must be coated with pure tin, done nowadays by electrolysis. This tinplate must be renewed regularly. Bowls for beating egg whites and sabayon saucepans are not coated, however, because whisks scrape the tin, which is quite a soft metal. Also, too much heat must be avoided with tinplated utensils to avoid melting the tin.

Why do I have a few misgivings about praising the physical nature of copper to you? Because, as a chemist, I suspect the state of the surface of the material your saucepan is made out of matters more than the nature of the metal itself. A porous copper would no doubt be disastrous. There are studies in progress. . . .

It remains a fact that copper is beautiful. And expensive as well. It can very adequately be replaced by another heat-conducting metal like aluminum, but the aluminum must be thick enough to prevent burning.

Why Use Wooden Spoons?

Wooden spoons are present in all kitchens. They benefit from the current taste for natural products, but they truly do deserve their place because they do not conduct heat. Left in a preparation that is cooking, they can be handled without burning the fingers of the cook. What a blessing that this tool, and the material of which it is made, wood, does not scratch the tin that lines the inside of our copper saucepans!

Mysteries of the Kitchen

Unanswered Questions

In this exploration of the wonderful world of the gourmand, we have had the opportunity to discover some answers. Nevertheless, cooking is teeming with questions. It is my dream that science will help us to answer them.

Here are just a few:

Supposedly, a sabayon can boil without turning if a pinch of flour is added to the mixture of egg yolk beaten into a liquid (water, wine, juice...). Experience shows that this precaution is effective. How does the flour act to protect the sauce?

If egg yolk is added to coarse sugar without being worked in, it cannot later be incorporated into the cream or the dough with which we want to mix it. The egg yolk "burns." Why is that?

When preparing a stock or a brown sauce, the ingredients are boiled for a long time in water. Which components escape with the water vapor (we smell them) and in what proportion? Which remain? How to influence this distribution?

Can a mixture of oil and butter be heated to a higher temperature than butter alone?

It is said that, when preparing a sauce, liquid can only be added to a roux when the saucepan is away from the heat. Why?

Apple juice turns dark. How can that be avoided?

Why does bouillon prepared in a saucepan covered with a lid become cloudy, and why must it be brought to a boil slowly?

Why is parsley used in short-term marinades (a day or two) but not recommended for longer ones?

Is it true that a suckling pig served at the table must have its head cut off immediately, or its skin will not be tender?

Why does excessive kneading make pie dough rubbery?

Why does a purée become rubbery if it is overworked or if it is worked at either too hot or too cold a temperature?

Why does adding a small quantity of liquid to mayonnaise whiten it as well as making it more fluid?

When preparing jam, does the kind of metal the saucepan is made out of matter?

Is it true that champagne will not make bubbles in glasses that are washed in a dishwasher?

If a little spoon is placed in the neck of an open champagne bottle, does that keep bubbles from escaping? If so, why?

Is it true that you can avoid releasing too many bubbles by first pouring a small amount of champagne into a glass before filling it completely?

Does the speed with which a marinade soaks into meat depend on the type of meat it is?

Can you make a successful aioli without egg yolk, using just garlic and oil?

Why does gelatin added to boiling milk make it turn?

Do you know the answers to these questions? If you would like to share them with me, I would be much obliged. Do you have other questions? Let me know what they are. I will try to find the answers.

And, in the meantime, *bon appétit!*

Glossary

A

AAAH!: The cry of delight guests utter when the first dish arrives. The sleight of hand responsible for the most beautiful "aaahs" cannot be explained in terms of physical chemistry.

ACETIC ACID: The main acid compound of vinegar.

ACID: Any substance that gives the impression of acidity; for chemists, these are molecules that, in solution, release hydrogen ions (H^+; hydrogen atoms that have lost their single electron). In cooking, the principal acid solutions are lemon juice and vinegar.

ACIDITY: A sensation communicated by substances like vinegar or lemon juice. Acidity is measured on the pH scale, from 0 to 14. Solutions with a pH lower than 7 are acid; solutions with a pH higher than 7 are basic.

ACTIN: One of the principal proteins in muscles, responsible for muscle contraction. When meat is cooked, the actin coagulates.

ALBUMINS: Small proteins soluble in water. Ovalbumin is one of these, present in egg whites, for example. In blood, there is serum albumin. The word "albumin" is generally used incorrectly in cookbooks, where it really means protein, a word that replaced "albumin" in chemistry about a century ago.

ALCOHOL: Any organic molecule with a carbon atom bound to an oxygen atom that is then bound to a hydrogen atom (-C-O-H). The most common alcohol, the one in wine, brandy, and liqueurs, is ethyl alcohol, with the formula CH_3CH_2OH.

AMINO ACIDS: In binding together like links in a chain, these molecules form proteins. The molecules of amino acids are characterized by the presence of a carbon atom to which are bound especially an acid group COOH (the letter C represents the carbon atom, O the oxygen atom, and H the hydrogen atom) and an amino group NH_2 (with a nitrogen atom [N], bound to two hydrogen atoms). Plant and animal organisms contain twenty types of amino acids.

AMYLASE: An enzyme that breaks down starch molecules.

AMYLOPECTIN: This is a polymer, that is, a molecule formed by the linking of many small identical molecules. The links in amylopectin are glucose molecules. The molecule is branched and insoluble in cold water.

AMYLOSE: Like amylopectin, but this polymer is in straight chains and soluble.

ASPARTAME: This is a sweetener, that is, a compound with a sweet taste. It dissociates in heat, releasing phenylalanine, which is bitter.

ATOM: A structure classically represented in the form of a nucleus around which rotate electrons. The nucleus is composed of protons, particles with a positive electrical charge, and neutrons, which are neutral. Having a negative electrical charge, the electrons are generally retained close to the nucleus by the forces of electrical attraction that are exerted between opposite charges.

AUTOXIDATION: A chemical reaction that produces rancidity in fats. It takes place rapidly in the presence of oxygen.

B

BÉARNAISE: One of the crown jewels of French cooking (I have a weakness for it; don't tell my wife!). A sauce composed of melted butter emulsified (*see* Emulsion) in a reduction of white wine, shallots, and vinegar. Egg yolks provide surface-active molecules for this emulsion, and their proteins coagulate, making microscopic lumps. Indeed, a successful béarnaise is a failure, microscopically.

BÉCHAMEL: A classic sauce made by diluting a roux (*which see*) with milk or bouillon.

BEURRE BLANC: Literally, "white butter"; a delicious sauce with fish. This is an emulsion obtained by stirring butter into a small quantity of liquid. It is a good idea to begin with cream.

BEURRE MANIÉ: Literally, "handled butter"; cold butter kneaded with flour, used as a thickener. Added to a sauce that is too thin, it provides the necessary viscosity. It is a stopgap measure, because the taste of raw flour is objectionable to true gastronomes.

BINDING: Or "thickening"; an operation meant to increase the viscosity of a sauce.

BISCUIT: Literally, "cooked twice"; "biscuit" is the French name for a sponge cake, different from a génoise sponge cake in that the egg whites are beaten into stiff peaks separately from the egg yolks and sugar.

BRAISING: A very gentle cooking process that enhances the taste of meat. A classic braising procedure consists of two stages: browning the meat by passing it through a very hot oven in order to "caramelize" the surface; then long cooking at temperatures lower than 100°C (212°F) to tenderize the meat without drying it out. Putting strips of meat, bacon, or ham around the meat that is being braised prevents the loss of juices.

BRINE: A solution containing more salt than can be dissolved in it. It is used in cooking for extracting through osmosis (*which see*) the water from plant and animal cells and thus preventing the proliferation of microorganisms.

BUTTER: Obtained by churning cream, this is an emulsion composed of small water droplets dispersed in milk fat. When you stir preparations that contains milk or cream, like mixtures for mousse, mousseline, or whipped cream, be careful to cool them in order to prevent them from turning into butter through cooling.

C

CAPILLARITY: Through the action of capillarity, water is introduced into very small spaces, like the interstices between starch granules in flour.

CASEIN: Eighty-five percent of the proteins in milk are casein. Casein molecules aggregate when the milk becomes acid or too salty: the milk curdles.

CATALYST: A molecule that prompts a chemical reaction.

CELLS: Vegetables, meats, the human organism—all are composed of billions of cells, microscopic sacks, each enclosing a structure called a nucleus, in a complex aqueous environment, the cytoplasm. All living cells are confined by a membrane. In addition, plant cells are protected by a rigid wall.

CHEMICAL REACTION: The process by which many molecules that encounter one another can exchange atoms and be rearranged.

CHEMISTRY: Among the most beautiful of the sciences, the one dealing with molecules as they react. Scientists often say of chemistry, "It's just cooking." What an honor!

CHOLESTEROL: This is a lipid (*which see*). It is accused of all sorts of evils because of the risk of coronaries when the blood contains too much of it, but the cholesterol in our food is not the direct source of blood cholesterol.

CLARIFY: This is to give limpidity and transparency to bouillon, a sauce, and the like.

COAGULATION: An aggregation of proteins provoked by heating or acidification, for example.

COLLAGEN: Collagen molecules form sheaths around the muscle cells in meat. Collagen is responsible for meat's toughness. When it is broken down, by heat in the presence of water, gelatin results.

COLLOID: A dispersion of solid, liquid, or gaseous particles in a continuous phase, either solid, liquid, or gas. Sauces obtained by diluting a roux (*which see*) in a liquid, milk or bouillon, are colloids.

CONCENTRATION: The proportion of a molecule in a system. Also, the name of the operation that increases this proportion.

CONDUCTION: Constantly in motion, molecules transmit their energy by colliding with one another. That is how heat is spread by conduction. In an oven, for example, the interior of a roast is heated through conduction.

CONVECTION: The circulation of a fluid that appears, for example, when it is heated from below. Less dense than the other layers, the lower ones rise, pushing the upper layers of fluid out of their way.

COPPER: A metal with red highlights that conducts heat very well, perfect for saucepans.

CREAM: An emulsion that forms naturally on the surface of milk when fat droplets gather and rise to the surface because they are less dense than the water. The cream we buy in grocery stores or supermarkets has generally been cultured with microorganisms that stabilize it but give it an acidity that the product drawn off the surface of milk does not possess.

CUL-DE-POULE: Literally, "a hen's backside"; this is a hemispherical copper bowl cooks use when beating egg whites. It is reserved for this use and cleaned with a clean rag soaked in vinegar or lemon juice.

D

DEGLAZING: An operation that consists of recuperating the sapid and odorant molecules in the bottom of a pan by adding a liquid like bouillon, meat juices, or wine.

DENATURATION: Changing the structure of proteins; in other words, the protein chain folds back differently over itself.

DIFFUSION: The movement of molecules. A drop of coloring deposited in a glass of water is diluted because the molecules of the coloring diffuse in the water.

DIHYDROGEN SULFIDE: A foul-smelling molecule composed of one sulfur atom and two hydrogen atoms. It is released when hard-boiled eggs are cooked for too long.

DISTILLATION: A process supposedly invented by the Iranian doctor Avicenna in about the year 1000, though, in reality, it was probably used long before then. Distillation dissociates mixtures through successive evaporation of the constituents. When wine is heated, the ethyl alcohol escapes as vapor at 78°C (172°F), while the water does not boil until it reaches 100°C (212°F).

DISULFIDE BRIDGE: A bond between two sulfur atoms. These form especially between the amino acids called cysteines.

E

EGG: Composed of three principal parts, the shell, the yolk, and the white. The yolk is composed of half water and half proteins and other surface-active molecules, such as lecithins. The white is a solution of proteins in water.

EMULSION: The dispersion of droplets of a liquid fatty substance in water or, conversely, of water in a liquid fatty substance (the structure depends on the respective proportions of water and fat). The stability is increased when the droplets are coated with surface-active molecules. If the proportions of an emulsion are altered, it can invert itself. In cooking, the result of such inversions is generally disastrous.

ENERGY: I have failed to find a good definition for this major concept in science, but it is always a good idea to analyze physical phenomena in terms of energy.

ENZYME: A protein possessing a catalytic action.

ETHYLENE: A gas that plays a part in the ripening of fruit. Certain fruits release more of it than others. That is the reason why bananas ripen quickly when they are put together in a fruit bowl with oranges.

F

FATTY ACID: A long organic molecule in which one carbon atom of the structure bears an acid group COOH, called the carboxyl group.

FERMENTATION: A controlled transformation of a food involving microorganisms, yeasts for bread, yeasts and bacteria for wine, lactobacillus for sauerkraut.

FLOCCULATION: The regathering of droplets initially dispersed in an emulsion. It is also a preliminary step before coalescence.

FLOUR: A product obtained by grinding grains of wheat, rye, oats, corn, and so on.

FRUCTOSE: A sugar with a chemical structure that includes six carbon atoms.

FRYING: An operation that consists of immersing foods into very hot fat.

G

GAS: The whole of molecules weakly bound to one another and moving around randomly within the entire volume available to them. Also, it is the punishment of those who eat indigestible food products.

GEL: A semisolid, three-dimensional network formed when a solution contains jelling molecules, that is, molecules capable of bonding to one another and to a great quantity of water.

GEL (OR JELL): As a verb, the formation of a gel, generally by lowering the temperature of a solution containing jelling molecules.

GELATIN: A substance with strong jelling properties obtained through the dissociation of collagen. When a gelatin solution cools, the gelatin molecules tend to bond to one another to form triple helixes, as in collagen.

GÉNOISE: A sponge cake obtained by beating a mixture of whole eggs and sugar for a long time. It takes longer and is more difficult for the eggs to form stiff peaks since the whites are not beaten separately.

GLAZE: For the physicist, this is a thin coating of ice; for the cook, it is the jelled mass obtained by reducing a stock (*which see*).

GLIADINS: Insoluble proteins in flour.

GLOBULINS: Soluble proteins in flour. The name comes from the way they are folded back on themselves in the shape of globules.

GLUCIDES: Or, more simply, sugars. Their old name, carbohydrates, was given to them because these molecules have an overall composition of one carbon atom for one oxygen and two hydrogen atoms. They react with proteins when heated to form molecules that have color or delight the nostrils with their aroma.

GLUCOSE: A sugar with a structure that includes six carbon atoms. This is the "fuel" that living cells burn.

GLUTEN: In the presence of water, flour proteins form an elastic network that we call gluten. Try this experiment: knead flour and water for a long time; then hold the dough you obtain under running water; what remains is an elastic, insoluble mass, the gluten.

GLUTENINS: Insoluble proteins in flour.

GLYCEROL: This is the glycerin that you can find in a drugstore. It is present in wines, giving them sweetness and smoothness.

GOURMAND: A glutton with self-control, and this is why gourmands are good representatives of the world of culture.

GOURMET: A specialist in wines.

H

HANGING PHEASANT: Contrary to what many believe, this is not a process of putrification, which would be dangerous for one's health. Hanging pheasant is like hanging venison. It must be done with an unplucked bird, which is suspended by its tail feathers for two to ten days, depending on weather conditions. It is said that the immortal Brillat-Savarin, author of *The Physiology of Taste* and adviser to the French Supreme Court, always kept hung birds in his pockets, to the great discomfort of his colleagues.

HOLLANDAISE: A sauce similar to béarnaise (*which see*) but differing from it because of the reduction, which contains no wine or shallots.

HYDROGEN: The first chemical element. Its atom is simply composed of a proton and an electron. Its ion, the hydrogen ion, is the proton deprived of its peripheral electron in the course of a chemical reaction. In solution, the hydrogen ion is surrounded by many water molecules; a solution in which it is abundant will be acid.

HYDROGEN BOND: A weak bond between a hydrogen atom and a neighboring atom, a donor of electrons (an oxygen atom, for example) in the same or another molecule.

HYDROLYSIS: A very important chemical process, as it generates amino acids when proteins are used as a substrate. These amino acids impart a wonderful taste to dishes.

HYDROPHILIC: A term used to describe a molecule that dissolves in water.

HYDROPHOBIC: A term used to describe a molecule that does not dissolve in water.

I

ION: An atom that has gained or lost electrons. In water, ions surround themselves with water molecules.

J

JAM: A flavorful gel (*which see*) always kept on the highest shelf in the pantry.

JELLY: A flavored gel.

L

LACTOBACILLUS: A single-cell organism that releases lactic acid. This bacterium is found in sauerkraut or in bread dough left to ferment naturally (sourdough).

LECITHIN: A surface-active molecule found especially in egg yolk, but which has cousins in all cell membranes of plant or animal tissues.

LEAVENING AGENTS: Unlike yeasts, these are not microorganisms but mixtures of chemical compounds, such as baking powder or baking soda, capable of releasing a gas (often carbon dioxide) that makes food preparations rise. They are also called leavening powders.

LIPID: From the Greek, *lipos*, fat. These molecules are defined by their insolubility in water. Food contains a wide variety of types of fat.

LIQUID: Formed when molecules make a whole less coherent than a solid but more coherent than a gas.

LUMPS: The disgrace of cooks.

M

MAILLARD REACTIONS: Chemical reactions basic to cooking, since they take place between the sugars and proteins found everywhere in food. They produce compounds with odorant and color properties, like those in the crust of bread, beer, the crisp browned surface of meat, and so on.

MALTASE: An enzyme that decomposes the sugar called maltose.

Margarine: A soft, fatty substance made from many other substances, often vegetable in nature. Gastronomes often fault it for not having the subtle flavor of our best butters.

MAYONNAISE: This is an emulsion (*which see*), or a dispersion of oil droplets in water, the latter having been provided by egg yolk and possibly by vinegar or lemon juice. No mustard in mayonnaise, otherwise it is no longer mayonnaise but remoulade (please do not confuse hammer and screwdriver, cats and dogs, science and technology, or mayonnaise and remoulade!)

MEAT: A muscular mass, composed of elongated cells, the muscle fibers, which sometimes reach twenty centimeters in length. Each fiber is wrapped in a sheath of collagenic tissue, and the sheathed fibers are gathered together in a bundle by other sheaths of collagen.

MERINGUE: A solidified foam obtained by baking stiffly beaten egg whites to which sugar has been added. Meringues must be baked gently at a low temperature.

MICELLE: A sphere formed by surface-active molecules; in water, for example, the hydrophobic tails of the surface-active molecules gather together, and the hydrophilic heads position themselves on the periphery, in contact with the water.

MICROWAVE: A wave similar to light, with a different wavelength. Microwaves are composed of an electrical field and a magnetic field; they prompt the alignment of molecules like water, where the distribution of the electrons is not uniform. Thus stimulated in one direction and then another in a very rapid rhythm, the water molecules become agitated and then agitate the surrounding molecules. This movement of molecules corresponds to an increase in temperature.

MOLECULE: An assemblage of atoms linked by chemical bonds. Molecules are formed and transformed by chemical reactions. They are not altered during physical transformations of matter.

MOUSSE: Or foam, the distribution of air bubbles in a solution or a solid. Egg whites beaten into stiff peaks form a liquid foam; meringue is a solid foam.

MYOGLOBIN: One of the proteins responsible for the color in meat.

MYOSIN: One of the principal proteins in muscles, responsible for muscular contraction. When meat is cooked, the myosin coagulates.

N

NITRATE: A salt in which one of the ions is nitrate, composed of one nitrogen atom and three oxygen atoms. Used in the salting process.

NITRITE: A salt in which one of the ions is nitrite, composed of one nitrogen atom and two oxygen atoms. Also used in the salting process.

O

OLIGOSACCHARIDE: A molecule composed of a number of monosaccharides. In other words, a small sugar composed of a few elementary sugars.

OSMAZOME: According to Brillat-Savarin, this was *the* taste "principle" in meat. The shortcomings of nineteenth-century chemical analysis misled the great gastronome. Meat's flavor results from the presence of a great number of molecules. As Valéry said, "anything that is simple is false."

OSMOSIS: A phenomenon that results from the uneven distribution of molecules. To explain briefly, a system made up of many kinds of molecules is in balance when the concentration of each type of molecule is identical in all parts of the system. If a salt crystal, for example, is deposited on the surface of a cell, the water molecules leave the cell so that their concentration will be equal in the cell and in the salt crystal.

OVALBUMIN: One of the proteins in egg white.

OXYGEN: This is the gas that our red corpuscles transport from the lungs to all our cells. In a water molecule, an oxygen atom is bound to two hydrogen atoms.

P

PAPAIN: One of the proteins present in fresh papaya juice. It reacts with other proteins by decomposing them.

PASTEURIZATION: The rapid heating of foods for the purpose of destroying the microorganisms that would spoil them.

PECTIN: A polymer present in plant cell walls. It forms the gel in jams.

pH: The unit of measure for acidity.

PHENOL: An organic molecule containing an aromatic ring of six carbon atoms, one of which is linked in particular to an alcohol (OH) group.

PHOSPHOLIPID: A lipid with one extremity that bears a hydrophilic phosphate group. Phospholipids are surfactants because of their hydrophilic part and their lipidic, hydrophobic part.

PHYSICS: The science of matter in general. Along with chemistry, it ought to be a help to cooks.

PIANO: A great chef's piano is his stovetop and work surface.

POLYMER: A large molecule formed by the linking of subunits called monomers. If you imagine a chain, the links would be the monomers.

PROTEIN: A chain in which the links are amino acids residues. In plant and animal organisms, these supple molecules fold over on themselves following specific patterns. By increasing the movements of atoms and various parts of the molecules, heat destroys the patterns natural to proteins. Thus we say that the proteins are denatured.

PUFF PASTRY: Obtained by six successive foldings of pastry dough into thirds. The result is 729 layers of dough separated by butter.

R

REDUCTION: The process by which, through heating, the excess liquid in a dish, a sauce, or a garnish is evaporated. Reduction is fundamental in cooking. Not only does it give a preparation its final viscosity, but it is also often vital to the development of flavor and aroma. This is where physics and chemistry merge, the height of culinary alchemy.

ROUX: A pastelike preparation obtained by cooking flour or starch in a fatty substance. Diluted, it thickens the aqueous solution that is added to it because it provides starch granules that swell and release bulky amylose and amylopectin molecules.

S

SABAYON: A delicious dessert obtained by mixing eggs (especially the yolks) and sugar and then adding an alcoholic liquid. Cooking it, after adding a pinch of flour, results in its thickening.

SALT: I pity those who lack it. Cooking salt is sodium chloride, which, in solid form, is composed of a network in which chloride ions and sodium ions alternate. Chemists also call salts those substances obtained by the reaction of an acid and a base (cooking salt can be obtained by the reaction of hydrochloric acid and soda).

SALTPETER: Potassium nitrate. This is an explosive, but it is very useful in the salting process.

SAUERKRAUT: A food obtained by fermenting cabbage in a brine (which see). Have you ever tried stuffed pheasant on a bed of fresh sauerkraut?

SKIMMING: The process by which a sauce is refined.

SOLID: A cluster of molecules very close to one another and immobilized by intermolecular forces.

SOLVENT: A liquid used to dissolve molecules. Lipids (which see) are good solvents of odorant molecules, as are terpenes (which see). Water is the main solvent of foods.

SOUFFLÉ: Has only one fault: it collapses.

STARCH: Granules made of two kinds of molecules, amylose and amylopectin (*see both*). Starch granules make gels when water diffuses into them.

STOCK: A concentration of flavors and gelatin that is obtained by browning fish or meat in a very hot oven and then cooking it for a long time in a large quantity of water in the presence of carrots, onions, and . . .

SUCROSE: This is table sugar, a disaccharide composed of glucose and fructose residues.

SUGAR: This is the crystallized form of the sucrose molecule. Sugar deposited on the surface of fruit or meat extracts the water from it by the phenomenon of osmosis. "Sugar" is also synonymous with "glucide."

SURFACTANT: A molecule composed of one part that dissolves easily in water and one part that is happier in fatty substances like oil. This kind of molecule can stabilize small oil droplets in water by coating the surface of these droplets, with the hydrophobic tail in the oil and the hydrophilic head in the water. Conversely, surfactants, also called surface-active molecules, can disperse drops of water in oil by arranging themselves with their heads in the water droplets and their tails in the continuous phase of oil.

SWEETENER: A compound that tastes like and is used as a substitute for sugar.

T

TANNIN: Tannin's astringency is due to its property of binding itself to the lubricating proteins in the saliva and blocking their functions. These molecules are phenolics.

TASTE BUDS: The group of cells on the tongue and in the mouth that possess receptors for sapid molecules.

TEMPERATURE: A value indicated by a thermometer, which must not leave the kitchen because its use is so helpful in cooking. The higher the temperature of a substance, the more the molecules in this substance are agitated by rapid, random movement.

TENDERNESS: A term used for meat; different from human tenderness but resembling it.

TERPENES: Odorant molecules from plants.

V

VANILLIN: The molecule principally responsible for the aroma of vanilla. Exactly the same molecule, with exactly the same atoms in the same positions, is found in vanilla beans and in the test tubes of chemists, but the synthesized one costs much less. This does not mean, however, that the odor of vanilla is the same as the odor of vanillin, as vanilla contains many other odorant molecules.

VINAIGRETTE: A fairly stable emulsion of oil in water. It lacks the surface-active molecules of egg yolks that would turn it into mayonnaise.

VISCOSITY: A fluid is viscous if it flows with difficulty. Certain sauces, such as béarnaise, have a viscosity that depends on their rate of flow. Very viscous when it is immobile, béarnaise sauce takes on a sublime fluidity when it enters the mouth. My mouth is watering at the very thought of it.

W

WATER: It is ubiquitous in food. There is the story of the oenologist who, while tasting wine with his eyes blindfolded, was given a glass of water without knowing it. "Hmm! This one doesn't have much odor or taste. I can't seem to identify it, but I can assure you that it won't sell."

Y

YEAST: A wonderful microorganism when domesticated to make bread, kugelhopf, . . .

Index